CW00969120

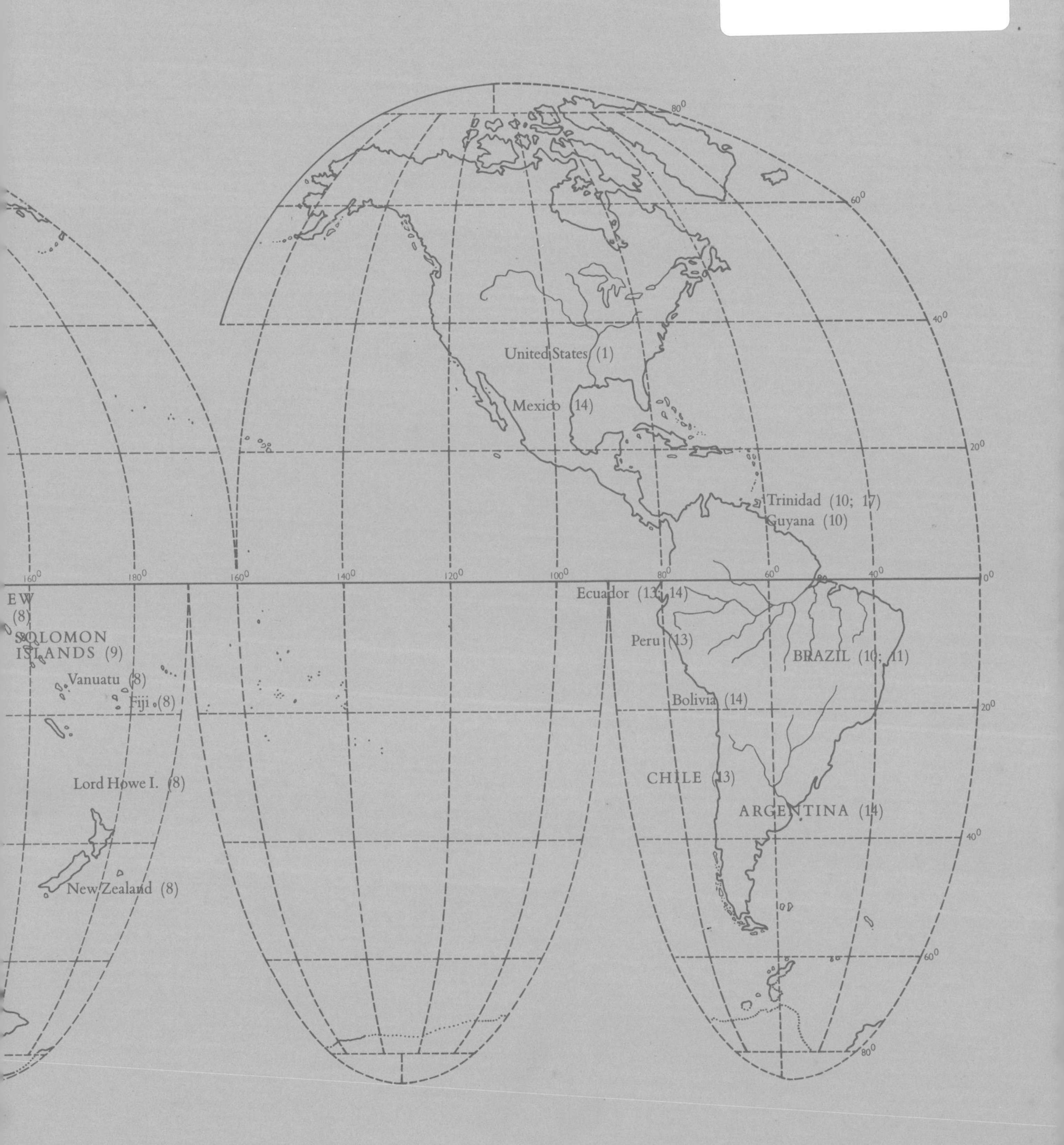
United States (1)
Mexico (14)
Trinidad (10; 17)
Guyana (10)
Ecuador (13; 14)
Peru (13)
BRAZIL (10; 11)
Bolivia (14)
CHILE (13)
ARGENTINA (14)
EW
(8)
SOLOMON
ISLANDS (9)
Vanuatu (8)
Fiji (8)
Lord Howe I. (8)
New Zealand (8)
160°
180°
160°
140°
120°
100°
80°
60°
40°
0°
80°
60°
40°
20°
20°
40°
60°
80°

Plant Hunting for Kew

Front cover
Base camp of a Kew expedition in August 1981 in the Banin Khola to the east of Mount Everest in Nepal. *Photo: C Grey-Wilson*

Back cover
Main picture: Nigel Hepper plant collecting with Benjamin Daramola of the Forest Herbarium, Ibadan, during the 1958 Kew expedition to Northern Nigeria.

Bottom left: *Dendrobium mohlianum*, New Georgia.

Frontispiece
Tropical montane forest in Kalimpong district, West Bengal, with an understorey of bamboos, photographed during a Kew Seed Bank collecting expedition to India in 1979. *Photo: R J Probert*

Half-title page
The forest coconut *Voanioala gerardii* being collected in Madagascar.

Title page
A climbing legume *Lonchocarpus glabrescens*; the roots are used as fish poison in Brazil.

Foreword by
SIR DAVID ATTENBOROUGH

Plant Hunting for Kew

Edited by
F NIGEL HEPPER

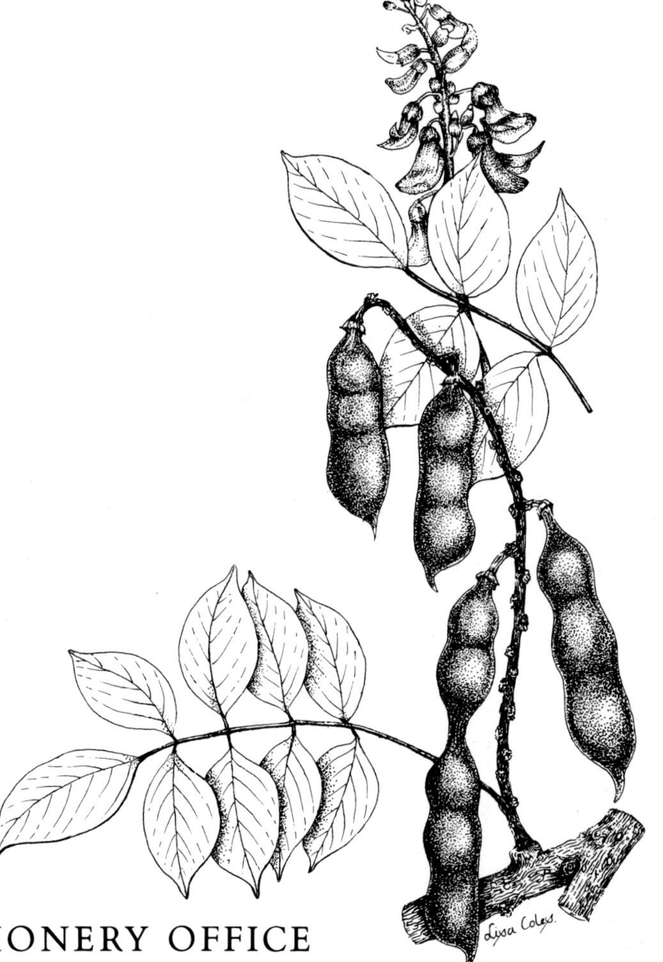

LONDON: HER MAJESTY'S STATIONERY OFFICE

First published 1989

British Library Cataloguing in Publication Data

A CIP catalogue record for this book is available from the British Library

HMSO publications are available from:

HMSO Publications Centre
(Mail and telephone orders only)
PO Box 276, London, SW8 5DT
Telephone orders 01–873 9090
General enquiries 01–873 0011
(queuing system in operation for both numbers)

HMSO Bookshops
49 High Holborn, London, WC1V 6HB   01–873 0011 (Counter service only)
258 Broad Street, Birmingham, B1 2HE   021–643 3740
Southey House, 33 Wine Street, Bristol, BS1 2BQ   (0272) 264306
9–21 Princess Street, Manchester, M60 8AS   061–834 7201
80 Chichester Street, Belfast, BT1 4JY   (0232) 238451
71 Lothian Road, Edinburgh, EH3 9AZ   031–228 4181

HMSO's Accredited Agents
(see Yellow Pages)

and through good booksellers

ISBN 0 11 250038 2 paperback edition

ISBN 0 11 250053 6 cased edition

Designed by HMSO Graphic Design

Printed in the United Kingdom for Her Majesty's Stationery Office
Dd 289378   C70   10/89

Contents

List of contributors viii

Foreword ix
Sir David Attenborough
Trustee of the Royal Botanic Gardens, Kew

Preface x
Professor Ghillean T Prance
Director of the Royal Botanic Gardens, Kew

Acknowledgements xii

Introduction xiii
F Nigel Hepper
Editor

PART I KEW EXPEDITIONS BEFORE THE SECOND WORLD WAR

1 **Strange and curious plants (1772–1820)** 1
Ray Desmond

2 **From rhododendrons to tropical herbs (1820–1939)** 11
Ray Desmond

PART II RECENT KEW EXPEDITIONS AROUND THE WORLD

3 **Tropical Africa continues to surprise** 23
Two treks in West Africa 23
F Nigel Hepper

When it rained in East Africa 31
R M Polhill

The Malawi connection 40
R K Brummitt

4 Searching for a forest coconut in Madagascar 51
John Dransfield

5 Tortoise islands in the Indian Ocean 61
S A Renvoize

6 Plant hunting in the Near East 72

The Wahiba Sands of Oman 72
T A Cope

Springtime in Cyprus 78
R D Meikle

7 Botanising in mountain Asia 85

To the Wakhan Corridor, Afghanistan 85
Christopher Grey-Wilson

Floral glories of Nepal Himalaya 93
A D Schilling

China's Yunnan mountains 96
Christopher Grey-Wilson & A D Schilling

8 Botanical exploration in Papua New Guinea 103
Martin J S Sands

9 Orchid hunting in the Solomon Islands 117
Phillip Cribb

10 From coast to campo in Bahia, Brazil 128
Simon J Mayo

PART III RECENT KEW SPECIALIST EXPEDITIONS

11 Quest for useful legumes 147
Gwilym Lewis

12 New horticultural plants 162
John Lonsdale

13 Travel and expeditions by Kew students 170
Leo Pemberton

14 Plant collecting for the Jodrell Laboratory 176
David F Cutler

15 Kew Seed Bank collecting 187
Simon Linington

16 Fern hunting in the tropics 196
Barbara Parris

17 Fungi: the Fifth Kingdom 203
David Pegler

18 Plant collecting and conservation 209
Grenville Lucas

Further Reading 214

Index 215

As well as pictures and maps illustrating the text in each chapter, several double pages of colour photographs extend the coverage to other expeditions. These appear on the following pages:

70–71 Indian Ocean islands expeditions
80–81 Near Eastern expeditions
124–5 Australasian and Pacific expeditions
140–41 South American expeditions

List of Contributors

Dr R K Brummitt (Herbarium) an authority on botanical nomenclature and the flora of Malawi.

Dr T A Cope (Herbarium) a specialist on the grasses of the Middle East.

Dr Phillip Cribb (Herbarium) an orchid specialist, particularly those of the Far East.

Dr D F Cutler (Jodrell Laboratory) Head of the plant anatomy section.

Ray Desmond formerly Chief Librarian and Archivist at Kew and then at the India Office Library.

Dr John Dransfield (Herbarium) a specialist on the palms of the world.

Dr Christopher Grey-Wilson (Herbarium) an authority on alpine plants and Editor of the *Kew Magazine*.

F Nigel Hepper (Herbarium) specialises in West African plants and the botanical exploration of tropical Africa.

Gwil P Lewis (Herbarium) a specialist on legumes, particularly those in South America.

Simon Linington (Wakehurst Place) Kew Seed Bank Manager.

John Lonsdale (Living Collections) Assistant Curator, Technical Section.

Grenville L Lucas Keeper of the Herbarium, Deputy Director and Chairman of Species Survival Commission IUCN.

Simon J Mayo (Herbarium) a systematic botanist interested in the Brazilian flora, especially the Araceae.

R D Meikle formerly a systematic botanist in the Herbarium; author of *The Flora of Cyprus*.

Dr Barbara Parris formerly specialising in tropical ferns at the Herbarium.

Dr David Pegler (Herbarium) Assistant Keeper and mycologist specialising in tropical fungi.

Leo Pemberton formerly Supervisor of Studies in the School of Horticulture.

Dr R M Polhill (Herbarium) Assistant Keeper, specialist on tropical African mistletoes and legumes.

Professor Ghillean T Prance Director of Kew and an authority on the Amazonian flora.

S A Renvoize (Herbarium) specialist on the grasses of South America and their anatomy.

Martin J S Sands (Herbarium) Coordinator and systematic botanist interested in the genus *Begonia*, SE Asia and Australia.

A D Schilling Deputy Curator, Wakehurst Place, and an authority on the plants of Nepal.

Foreword

The heroic days of plant collecting are famous and the name of Kew echoes throughout them. The transfer of the rubber tree from Brazil to Malaya, the discovery and cultivation of quinine, the collection of rhododendrons from the flanks of the Himalayas and their establishment in gardens from one end of the earth to the other, by such achievements in the last century, Kew and its collectors enhanced the health, the wealth and the beauty of the world.

The enterprise that brought such rewards still flourishes at Kew. What is more, it is more needed now than ever before. How paradoxical it is that at a time when we are exploring the moon and searching for evidence of life in galaxies light-years distant, we still do not know all the plants that grow on earth. A century ago, Kew's botanists voluntarily took on the task of classifying some of the most important of all plant families and listing every species of plant growing in particular regions of the world. They brought back specimens for the Herbarium and seeds for germination in the Gardens. Their printed works are massive monuments to their scholarship and the living plants themselves, flowering and seeding in spite of the vagaries of London's climate, in the Gardens beside the Thames, marvellous evidence of horticultural skill.

But the work is far from being finished and it is more arduous than you might suppose. The forests and the deserts, the swamps and the mountain tops where the collecting must be done are much the same as they ever were. Here are the same mud and the same mosquitoes, the same searing heat and the same drenching rain. What has changed is the urgency of the task. Not so long ago, the tropical forests seemed eternal and there was all the time that anyone needed to explore them. Today, those forests are disappearing before our eyes as other human beings, carelessly or deliberately, destroy them. What quinine and rubber, what food and what beauty are we losing? Only plant collectors can tell us of such impending disasters and only botanic gardens prevent them.

The heroic days are by no means over. Here is a chance to savour them while they are still dawning.

David Attenborough

Preface

The great botanic gardens of the world have always been in the forefront of plant exploration, and the Royal Botanic Gardens, Kew, is no exception. Kew's standing as a great botanic garden derives from the excellence of its collections. Over 10 per cent of the world's flora (50,000 species) is represented in the living collections at Kew, which also has one of the largest herbaria in the world, an outstanding library of botanical books, paintings and drawings and an unrivalled collection of botanical artefacts in its economic botany museums.

These fine collections are the result of the continual emphasis placed on plant-hunting expeditions since the middle of the 18th century. It is because of these expeditions that we have the range of exotic plant species that we grow in the gardens and the herbarium specimens on which to study the classification and evolutionary systems of the plant kingdom. Equally, these expeditions and the resulting research have led ultimately to the introduction of many plant products that we use in our daily life.

This volume shows that the Royal Botanic Gardens has played, and continues to play, a major role in all three aspects of plant hunting. The quest for new plants and plant information has not ceased. The gardens' staff are still active in many parts of the world and are still finding many new species. At a time when we have explored the surface of the Moon and sent rockets to other planets, there is a great deal yet to be discovered about the plants of our own planet. The tragedy is that the environmental degradation of this century is causing many plants to become extinct before they have even been catalogued. The urgency of further plant exploration cannot be overemphasised.

The descriptions given here of plant hunting are varied: most contain the basic facts in review, many are very personal accounts and one presents extracts from a contemporary journal. All are expertly woven together by the skill of the editor, F Nigel Hepper. The record is impressive but you should remember, as you read, that what is not described in detail are the many hardships that plant hunters have undergone to achieve such results. The expeditions of a hundred years ago endured many health hazards and transport difficulties that do not occur today. However, even as I write this in 1989, I have just heard that the camp of a Kew expedition to Brazil has been completely washed away by a flash flood.

This book is a tribute to the many people who, despite many hardships, have served the cause of botany and horticulture as plant hunters for Kew. They can be proud of their part in the building of a great botanic garden with one of the most outstanding plant collections in the world. It is my hope that, as you read this book, you will enjoy travelling all over the world with them.

GHILLEAN T PRANCE
Director of the Royal Botanic Gardens, Kew

Acknowledgements

The editor and contributors gratefully acknowledge the assistance received from many colleagues, and others, in the preparation of this volume, and from all those who provided the illustrations. For photographic work we would also like to thank Andrew McRobb and Media Resources, and for the maps, Bryan Poole.

Introduction

The casual visitor to Kew who sees the wealth of living plants in the Palm House and Princess of Wales Conservatory, as well as in the Rock Garden and Arboretum, finds it hard to believe that many of them have been collected by Kew expeditions. The millions of dried specimens in Kew Herbarium also represent a lot of hard work and anguish, even risk of life, on the part of those involved in their collection. Some visitors are surprised that Kew botanists are still studying rain forests and deserts, as well as tropical swamps and equatorial mountains, in their attempts to understand the plant resources of the world.

In planning this book on Kew expeditions I wanted to make the accounts more or less contemporary, so I asked colleagues to write about their own plant-hunting safaris during the last few decades. True, there were famous Kew expeditions during the preceding two centuries, but as these have often been related I decided to condense them into a couple of chapters. These have been admirably written by Kew's former librarian, Ray Desmond and appear in Part I.

Since it would be impossible to find space for every recent expedition, a selection had to be made. Those chosen fall into two main categories – those with a geographical emphasis and those of a more specialist botanical nature. Part II thus includes examples of general expeditions by Kew staff to places around the world, supplemented by double-page spreads of photographs which extend coverage of some chapters to take in other expeditions in the same geographical area. Part III includes specialist botanical expeditions, based on subjects such as horticultural plants and Jodrell Laboratory collecting, as well as ferns and fungi. The concluding chapter is a salutary reminder that we all depend on plant life and increasingly plant life depends on us. The more we know about it the better we can conserve it. Judicious scientific plant-hunting is not threatening the world's flora but is helping to protect it – and Kew's multilateral research is playing an important part in this movement.

Contributors have written for both lay and specialist readers who are interested in plants, and the accounts include personal recollections, expedition incidents and geographical descriptions, illustrated by photographs taken on the spot and plant portraits from *Botanical [now Kew] Magazine* and elsewhere. It is hoped that these articles with their varying styles will enthuse readers with the interest and excitement of botanical expeditions. However, there are many joys that are difficult to convey, such as seeing for the first time a huge tree hitherto only known from twigs on a herbarium sheet, or the smell of a tropical rainforest with birds calling far up in the canopy, or the colour of alpines on equatorial mountain slopes, or the sun rising over desert scrub as far as the eye can see. Kew botanists appreciate that plant hunting in remote corners of the Earth is a real privilege.

Planning such an expedition is a formidable undertaking. Cooperation with national administrators and local scientists is the key to the success of any expedition. But this also has the advantage to the nationals in that they receive training and experience in fieldwork, as well as helping to obtain the obligatory duplicates of plant specimens for their own botanical institutions. Most nations have stringent rules for the collection of any plant material for which permits are required since plants are a valuable resource which must be safeguarded. Kew's roles as a quarantine inspection-station and international centre for plant conservation are complementary to its other scientific functions.

Preparing for a botanical safari far from centres of population entails equipping a vehicle with sufficient stores for the duration of the trip. Although vegetables, bread and petrol may be available locally, the basic stores need to be purchased in advance. In any case enough plant presses, drying papers, bottles for material in pickle and bags for fruits and seeds need to be included as it is unlikely that any replenishments will be available. On the roof go containers for water and reserve petrol for when supplies are unreliable or distances between stops are long. Spare tyres and other vehicle parts may be necessary, while the occupants need suitable clothing, first-aid boxes, cooking stoves and tents. Sufficient funds in small denominations have to be taken since it is unlikely that up-country stores will accept a cheque or villagers be able to provide change for a high-value note. These are some of the basic requirements for any expedition, but each will have its own special needs and large and multidisciplinary expeditions present their own organisational problems.

Once out in the field or forest the botanist is faced with the fact that, unfortunately, the world's vegetation has been so disturbed and degraded by man's activities that fieldworkers need to search carefully for natural habitats and the wild plants of any locality. However, Kew expeditions are planned with a particular objective, such as studying a specific plant family or the plants of a certain habitat, and the specialist is able to make more discerning collections than a less informed person would do.

The collection of herbarium material is a laborious business since each gathering needs to be made with several duplicates, each collection must be given a number and detailed notes prepared. Portable presses are taken into the field for drying at night. Alternatively large polythene bags are carried, into which the plants are placed between flat newspapers and soused with industrial spirit to preserve the material until it can be pressed and dried at base camp. Living plants intended for cultivation also present problems for the collector: seeds, fruits, bulbs and succulents are easier to handle than whole herbaceous plants and cuttings, but skilful propagators can encourage the most unlikely material. The transport and storage of dried specimens are continual challenges to the depredations of termites, moulds and customs officials, so the thrill of receiving the material safe and sound at Kew can only be truly appreciated by the plant hunters themselves. If this book recalls even a fraction of that excitement, we will be well pleased.

F Nigel Hepper
Kew Herbarium

A geographical and physical globe of the world made by A. Keith Johnston FRGS for the 1851 London Exhibition.

Photo: Royal Geographical Society

PART I

KEW EXPEDITIONS BEFORE THE SECOND WORLD WAR

1 Strange and curious plants (1772–1820)

RAY DESMOND

When Sir William Chambers landscaped the grounds of Princess Augusta's estate at Kew during 1757–63, and embellished them with such well-known architectural conceits as the pagoda and the ruined arch, he paid tribute to Lord Bute, who advised the Dowager Princess on the planting, and also to 'the assiduity with which all curious productions are collected from every part of the globe', forecasting 'that, in a few years, this will be the amplest and best collection of curious plants in Europe'.[1] John Hill listed over 3000 species of plants in cultivation at Kew in his *Hortus Kewensis* (1768). By the middle of the 18th century the traffic in imported plants and seeds was substantial and the demand from enthusiastic botanists and gardeners insatiable.

It has been estimated that despite considerable losses through the hazards of long sea voyages some 500 species of foreign hardy trees and shrubs were successfully introduced into British gardens by the close of the 18th century. Many came from North America and were subjected to a comparatively short sea crossing. American plant collectors such as John and William Bartram found it impossible to satisfy the British market. In August 1761 John Bartram complained petulantly to Peter Collinson, who had recruited a syndicate of subscribers for new American plants, 'I have sent them seeds of almost every tree and shrub from Nova Scotia to Carolina; very few are wanting, and from the sea across the continent to the Lakes . . . if I die a martyr to Botany, God's will be done'.[2]

In William Aiton's *Hortus Kewensis* (1789) (Sir) Joseph Banks (1743–1820) is frequently acknowledged as being instrumental in introducing many of the plants listed. For Banks, who had been naturalist on Captain Cook's *Endeavour* voyage to the Pacific, botany was to be a compelling interest all his life. Befriended by George III, he advised the monarch on the royal gardens at Kew with an enthusiasm that confirmed his own personal commitment. He used his influence as president of the Royal Society to obtain plants from a global network of correspondents. The botanic gardens at Calcutta and Jamaica were

Francis Masson (1741–1805) was the first plant collector to be sent out on behalf of the Royal Botanic Gardens at Kew. He made notable collections in South Africa, including Cape heaths (*Erica* species) and stapelias (*Stapelia* species).

generous donors and Banks persuaded ships' captains to bring the plants to England.

It was his ambition to make Kew's collection of plants unrivalled and he jealously resisted the competing claims of other European botanical gardens. By 1789 he was describing the royal gardens as 'His Majesty's Botanic Garden of Kew', always vigorously exploiting the royal connection to acquire yet more plants. In 1797 he reminded the governor of New South Wales that 'Kew Gardens is the first in Europe, and that its Royal Master and Mistress never fail to receive personal satisfaction from every plant introduced there from foreign parts when it comes to perfection'.[3] It has been calculated that during the reign of George III some 7000 new species were introduced from abroad and Sir Joseph Banks played a dominant role in this migration of the world's vegetation.

Banks's active participation began in 1772 when Sir John Pringle, the king's physician and newly elected president of the Royal Society, suggested to George III that a Kew gardener, Francis Masson (1741–1805), should be sent to the Cape of Good Hope to collect plants for the royal gardens. The idea may have been Sir John's but its implementation was given to Banks.

Masson spent two and a half years travelling in a trek waggon or on horse, penetrating some 900 km (500 miles) into the interior, before returning in 1775. His collections, boasted a delighted Banks, enhanced Kew's 'superiority which it now holds over every similar Establishment in Europe'.[4] Masson reckoned he had added some 400 new species to the royal gardens. His cycad, *Encephalartos longifolius*, still survives in the Palm House.

After a plant-collecting expedition to the West Indies via the Canary Islands, the Azores and Maderia, Masson returned to South Africa in 1786, remaining there for nearly 10 years. A special 'Cape-house' had to be built at Kew in 1792 to house his introductions: ixias, mesembryanthemums, oxalis, pelargoniums and, in particular, heaths, which were represented by more than 80 different species. Masson died in Montreal in 1805, still collecting for Kew.

While he was on his second tour of duty in South Africa, Masson met another Kew collector, Anton Pantaleon Hove (*fl.* 1780s–1820s), who was passing through Cape Town on his way back from India. Polish by birth, Hove had already collected on the African coast when he was engaged in 1787 by Banks. He was despatched to India with instructions not 'to neglect those [plants] which are small or unsightly, as it is . . . likely that qualities useful to physic or manufacture, and, of singularity of structure interesting to the botanist, should be found in the minute and ugly, as in the conspicuous and beautiful productions of nature'.[5] The victim of several robberies by tribesmen, in which he lost not only his personal possessions but also his specimens, Hove's collections were consequently meagre in relation to his expenses; he was ordered by a disappointed Banks to return to England at once.

Banks's directive to Hove to seek out plants 'useful to physic or manufacture' was indicative of his constant interest in economic

The oldest plant under glass at Kew is this cycad *Encephalartos longifolius*, which was collected in Natal by Francis Masson in 1775. For the rebuilding of the Palm House it had to be moved in 1984 to a temporary glasshouse and is seen here with supporting scaffolding.
Photo: S Minter

exploitation. When planters in the West Indies petitioned the king to introduce the breadfruit from the Pacific as a cheap food source for their slaves, it was inevitable that Banks should be consulted about the proposal. He chose a Kew gardener, David Nelson, to collect and tend the precious cargo of breadfruit trees on board H M S *Bounty* under Captain Bligh.

Nelson had served as a midshipman with botanical duties on Captain Cook's third voyage (1776–80). His collection of plants and seeds impressed Banks, who saw him as a dependable collector. Nevertheless Banks solemnly reminded him that 'one day or even one hour's negligence may at any period be the means of destroying all the trees and plants which may have been collected; and from such a cause the whole of the undertaking will prove not only useless to the public, but also to yourself'.[6] What transpired is one of the famous episodes in British nautical history. At the time of the celebrated mutiny, Nelson remained loyal to Bligh and died on the island of Timor in 1789.

Bligh returned safely to England to command another ship, H M S *Providence*, which did successfully accomplish the mission of transporting the breadfruit trees to the West Indies. James Wiles and Christopher Smith, charged to look after the trees, were meticulously briefed by Banks. While the primary objective of collecting the breadfruit was always to be paramount, if, however, they discovered plants 'particularly beautiful or curious, you are to acquaint the commanding officer, who, if he thinks proper, will give you leave to take on board one of two of each sort for the use of His Majesty's Botanic Garden at Kew'.[7] Smith returned to Kew in 1793 with a varied cargo of plants from Tahiti, New Guinea, St Helena, St Vincent and Jamaica. While Smith was briefly superintendent of the botanic garden at Penang he still remembered Kew, which was the grateful recipient of several consignments of tropical plants.

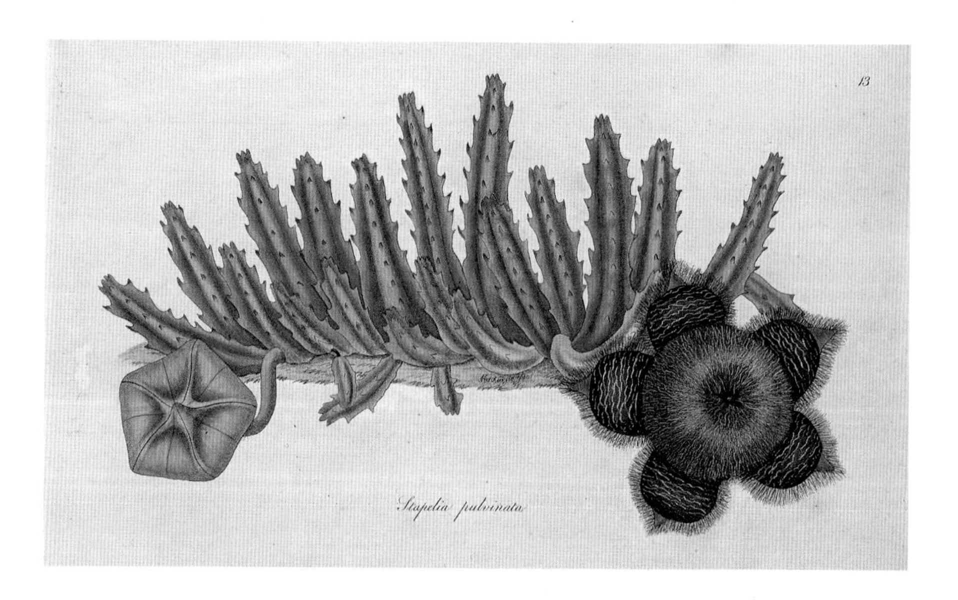

Stapelia pulvinata was collected in South Africa, drawn by Francis Masson and published in his *Stapeliae novae* (1796).

Peter Good, a young Scottish gardener, brought to Kew from Calcutta in 1796 one of Christopher Smith's collections. It would appear that he made another similar journey to Calcutta in 1799. When Matthew Flinders, one of Bligh's officers on *Providence*, proposed in 1800 an exploration of Australia's coastline, it was Banks's support he sought. The Admiralty needed little persuasion and equipped H M S *Investigator*, installing at Banks's instigation a plant cabin on the quarterdeck for storing living plants for Kew. Robert Brown was appointed ship's naturalist and Peter Good his assistant and gardener.

The *Investigator* reached Australia in December 1801 and soon packets of seeds were on their way to Kew. Banks informed Brown that 'they are all sown in Kew Gardens, & much hopes built on the success of them, which we expect will create a new epoch in the prosperity of that magnificent establishment by the introduction of so large a number of new plants as will certainly be obtained from them'.[8] After a voyage to Timor Peter Good died of dysentery at Sydney in 1803. Some 116 Australian species are recorded in the second edition of *Hortus Kewensis* (1810–13) and Richard Salisbury's naming an Australian genus *Goodia* was a commemorative gesture thoroughly deserved.

David Burton clearly knew of Sir Joseph Banks's devotion to Australian affairs when he wrote to him while on his way in 1790 to Port Jackson as superintendent of convicts and public gardener. He successfully requested the additional appointment as plant collector for Kew, being granted £20 a year provided he collected for no other person. Before his accidental death just a little over six months after his arrival in the colony he had despatched a few plants to Kew.

In 1784 Banks received a present of seeds from Archibald Menzies (1754–1842), a young naval surgeon serving in Canada. Five years later, through Banks's intervention, Menzies obtained the post of surgeon and naturalist on H M S *Discovery*, commanded by Captain Vancouver. Banks briefed him fully. He was to list all the vegetation he encountered, keep a record of soil characteristics and the climate and

Archibald Menzies, who in 1789 joined Captain Vancouver's voyage to Chile and western North America, sent back to Sir Joseph Banks many seeds of trees including the monkey puzzle *Araucaria araucana*, which was grown at Kew for the first time: one of them survived at Kew until the end of the 19th century.

consider whether the land would be suitable for English settlers. 'When you meet with curious or valuable plants which you do not think likely to be propagated from seeds in His Majesty's garden, you are to dig up proper specimens of them, endeavour to preseve them alive 'till your return.'[9]

Discovery sailed on its survey of the north-western coast of America in 1791 with Banks's plant cabin glistening on the quarterdeck. That floating glasshouse was one of the causes of friction between the autocratic and irascible Vancouver and the conscientious Menzies. While the ship was in Monterey Bay off the Californian coast. Menzies complained bitterly to Banks that 'goats-dogs-cats-pigeons-poultry etc. etc. are ever creeping in and destroying the plants' to the complete indifference of the captain. When the ship was nearing the end of its voyage in 1795 Menzies informed Banks that he was under arrest. Vancouver had commandeered the services of Menzies's servant, who had been engaged to look after 'those live plants I was directed to bring home for His Majesty's Royal Gardens', and these plants 'industriously collected in distant regions of the world' were now 'dead stumps'.[10] Menzies protested to Vancouver, 'who immediately flew in a rage, and his passionate behaviour and abusive language on the occasion prevented any further explanation'.

A watercolour of *Gentiana douglasiana* by Archibald Menzies, although named after a later collector, David Douglas. Menzies's dried specimen is in the Kew Herbarium and another is at the British Museum (Natural History).

Despite this contretemps the voyage amply repaid Banks's trust in Menzies, who harvested the seeds of many splendid trees – *Arbutus menziesii* for example, probably the finest evergreen in Northwest America, and that Victorian favourite, the monkey-puzzle tree, *Araucaria araucana*, collected in Chile.

Another of Banks's protégés was George Caley (1770–1829), the son of a farrier near Manchester. A diligent student of the British flora, he asked Banks's assistance in obtaining employment in the botanical world. 'I do not know that there is any trade by which less money has been got than that of botany', was Banks's discouraging reply. 'Except Mr Masson who travels for the King and who is now resident at the Cape of Good Hope, I know of no-one who by success in the practice of botany has gained anything.'[11] Sir Joseph (he was knighted in 1795) thought that a knowledge of the world's exotic flora was desirable.

Caley, obedient to this advice, obtained employment at the Chelsea Physic Garden and William Curtis's Brompton nursery. He spent a few months at Kew in 1798 but left because of poor pay and few prospects of employment overseas. Sir Joseph reprimanded him. 'You seem to have so good an opinion of the value of your own abilities as to think I ought to demand of Government to send you out [to Australia]. Suppose I was to agree to this opinion of yours, and a competent salary was to be allotted to you, pray tell me what you would propose to do for your employers in return.'[12] Nevertheless, when Sir Joseph learnt that the governor of New South Wales was returning to Australia, he got Caley a passage and agreed to support him in return for specimens for his own herbarium and seeds for Kew.

In a remarkably short time after landing at Sydney in 1800, Caley's impetuous and quick-tempered manner antagonised many of the colonists. When Governor King complained about his disruptive behaviour, Sir Joseph expressed sympathy. 'I feel a particular obligation to you for bearing with the effusions of his ill-judging spirit. Had he been born a gentlemen he would have been shot long ago in a duel.'[13] Caley published nothing of his botanical discoveries in New South Wales but Robert Brown, the eminent contemporary authority on the Australian flora, considered him to be 'a most assiduous and accurate botanist'.

Since the vegetation of Argentina was poorly represented in the royal gardens, Sir Joseph approached the Foreign Secretary, Lord Hawkesbury, in 1802 for permission to send a Kew gardener to Buenos Aires. Presumably the political tensions in Europe prevented its implementation. This disappointment for Sir Joseph was assuaged by an opportunity the following year to send another gardener, William Kerr, to China, where its xenophobic rulers confined Europeans to Canton and Macao.

David Lance, who was about to sail to Canton, where he was superintendent of the East India Company Factory, was willing to take a Kew gardener with him. William Kerr (d. 1814) was informed by Sir Joseph that he had been chosen 'thereby holding out to you a prospect, in case you are diligent, attentive and frugal, of raising yourself to a better station in life than your former prospects permitted you to expect'.[14] He was to view this appointment as botanical collector as an apprenticeship which David Lance would supervise.

Banks expected brief accounts of his progress in China; W T Aiton was to receive the fuller report, which was to include lists of the plants he had collected and despatched, and relevant data on climate, soils and horticultural practices of the Chinese. Sir Joseph had specific queries: for instance, how did the Chinese grow dwarf varieties, and what plants did they use in the manufacture of white rope. Kerr was advised that he could learn a great deal about the identification of the local flora by studying the flowers painted by Chinese artists on paper and furniture.

When Kerr reached Canton in 1803 he found that the limitations of movement imposed on all foreigners frustrated his attempts to collect. He had to content himself with the purchase of cultivated plants – azaleas, chrysanthemums, camellias and peonies – at the Fa-tee nurseries in Canton. Realising he was wasting valuable time, he went to the Philippines in 1805, returning to China with a fine collection of plants, most of which were unfortunately lost when a hurricane struck the ship. Carefully packed in boxes, chests and plant cabins, Kerr's acquisitions were shipped to London but few were alive by the end of the voyage.

Banks recommended that a Chinese gardener should accompany each shipment; 'he will be received at Kew where he may easily by employing himself earn a comfortable livelihood and much useful knowledge' and, Sir Joseph promised, 'he will be comfortably and properly maintained and sent back with the return cabbin the next season'.[15] Kerr's notable introductions included *Lonicera japonica*, a

Japanese plant much admired by the Chinese; the Heavenly Bamboo, *Nandina domestica; Rosa banksiae*; and the Tiger Lily, *Lilium lancifolium*, which W T Aiton propagated with such success that he distributed over 10,000 bulbs between 1804 and 1812.

Sir Joseph, pleased with Kerr's performance, nominated him for the post of superintendent and chief gardener at the Royal Botanic Garden in Ceylon (now Sri Lanka). William Kerr had at last achieved the 'better station in life' promised to the diligent collector. But he died in Ceylon in 1814, having enjoyed his promotion and good fortune for two years.

The need for ships to sail in convoys during the Napoleonic wars inevitably protracted voyages, thus exacerbating the difficulties of keeping the plants alive. Kerr was therefore the last gardener employed by Kew until the cessation of hostilities in 1814, when W T Aiton, the director at Kew, lost no time in seeking Sir Joseph's assent to send out collectors once more. 'I have in view men of sound principles and invaluable zeal for the service, having the best requisites of knowledge and desire to offer themselves as collectors, and who will perform this duty in any part of the world.'[16] Sir Joseph, anxious that Kew should not fall behind the German royal gardens at Schönbrunn which he considered to be its only serious rival, readily agreed. He selected the Cape of Good Hope and New South Wales for botanical exploration and still cherished a desire to send a man to Buenos Aires 'but until Spain has repossessed herself of her refractory colonies, this cannot be thought of'.[17]

With his customary persuasive skill he submitted a well-argued case to Sir George Harrison, assistant secretary at the Treasury. Concerned that the Treasury should not have any excuse to reject the scheme on the grounds of expense, he proposed various economies, stressing that the 'collectors must be directed, by their instructions, not to take upon themselves the character of gentlemen, but to establish themselves in point of board and lodging as servants ought to do'.[18] If approval were granted, he nominated a Kew gardener, Allan Cunningham (1791–1839), and a former Kew employee, James Bowie (*c.* 1789–1869) as plant collectors. The Treasury not only approved but were most cooperative; they arranged a passage for both men on a naval ship and even requested the governor of the Cape of Good Hope to supply a waggon, teams of oxen, a native driver and a Boer escort if the collector travelled outside the colony.

Cunningham and Bowie embarked in October 1814 for Brazil, where they were to seek a passage to the Cape. In the event they were in South America for two years, collecting around Rio de Janeiro, the nearby Organ Mountains and São Paulo some 400 miles distant. Unfortunately only a fraction of their collections, which included orchids and bromeliads, survived the voyage and were successfully cultivated at Kew. In September 1816 the two men parted company: Bowie sailed to the Cape and Cunningham joined a convict ship bound for Australia.

The confidence placed in them by Banks and Aiton was soon rewarded by the arrival at Kew of their botanical discoveries. Sir Joseph

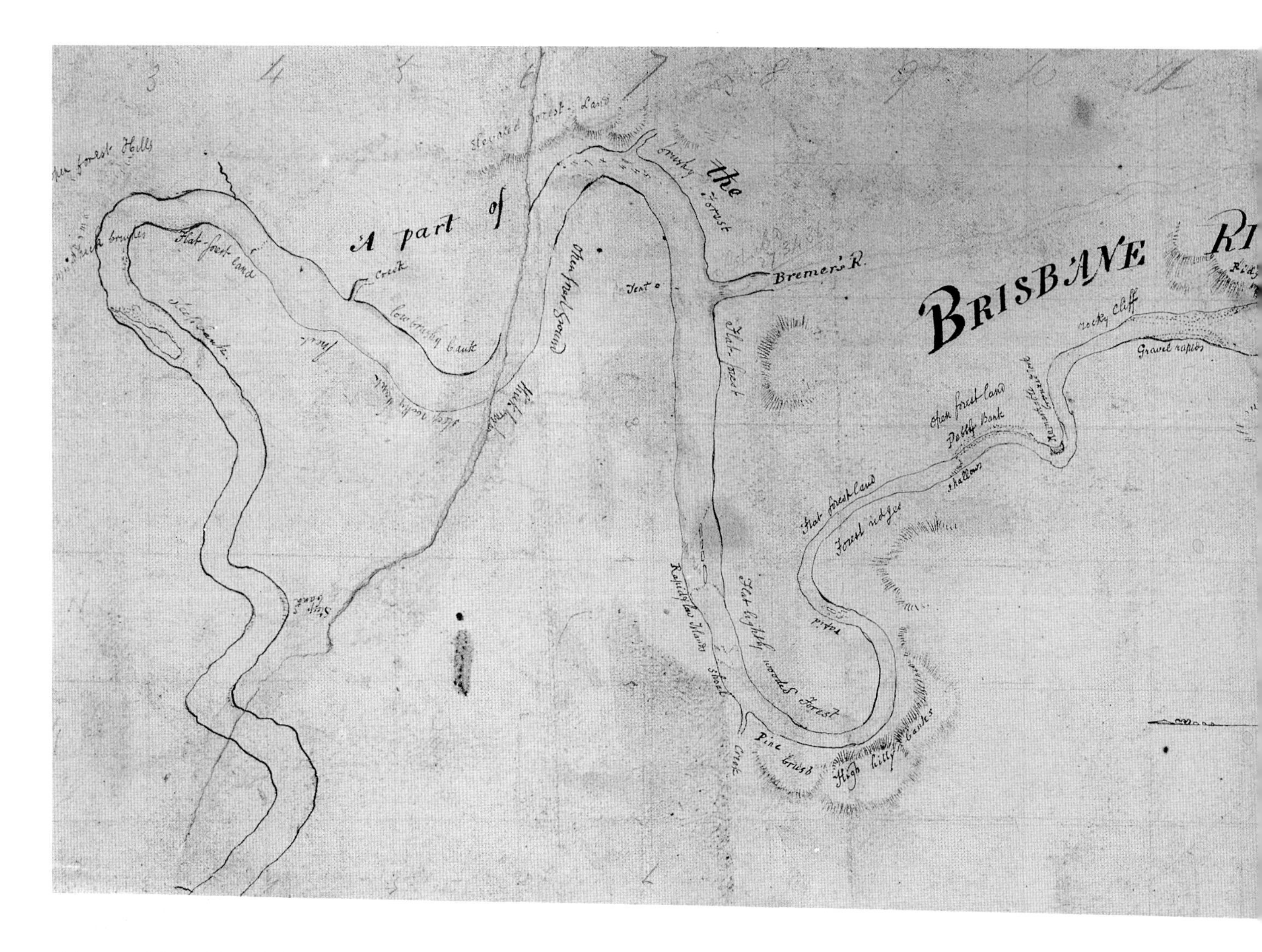

This map of the Brisbane River, in Queensland, was drawn by Allan Cunningham who reached Australia in 1817 and sent back to Kew many living plants. He was an early explorer and a competent surveyor, in addition to being a plant collector, and his maps are among the earliest cartographic records of eastern Australia.

congratulated Cunningham. 'I think Kew will be in as high a state of beauty and scientific excellence as it ever was when Masson sent home the beautiful and curious novelties as they then were from the Cape.'[19]

Although Kew had appointed a second foreman – the redoubtable John Smith – in 1823 to cope with the flood of Bowie's and Cunningham's introductions, Bowie's activities were terminated the same year, following a reduction in the parliamentary vote for the royal gardens. William Jackson Hooker, then Professor of Botany at Glasgow University, protested that 'this indefatigable naturalist, after sending the greatest treasures, both of living and dried plants to the Royal Gardens, and in the midst of his usefulness, has, by a needless stretch of parsimony, been recalled'.[20]

Within a few months of his arrival in Australia, Cunningham found himself the 'King's botanist' on the surveyor-general's exploration of the Macquarie and Lachlan rivers. Almost immediately on his return to Sydney with plant specimens and seeds, he was detailed to join an official survey of the Australian coast. Four subsequent voyages as

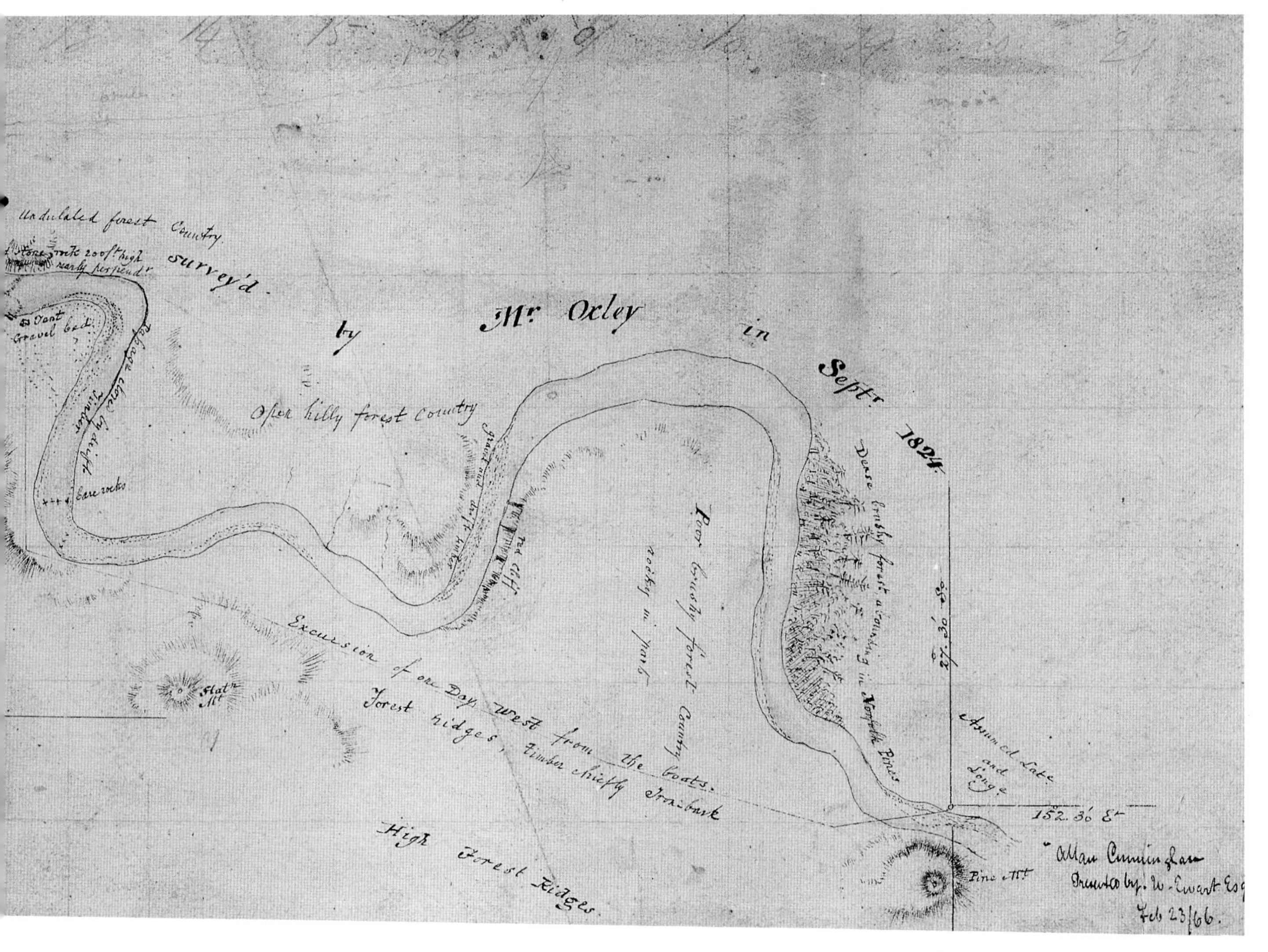

ship's botanist took him as far north as Timor and west into the Indian Ocean to Mauritius. In 1823 he found a pass through the formidable Liverpool Range of mountains to the rich grazing lands beyond. Three years later he became the first Kew collector to visit New Zealand. His reputation as an explorer was firmly established with his discovery of the Darling Downs, ideal for settlement and agriculture. On his travels he always took a bag of peach stones and seeds of other fruits and trees, planting them wherever he thought they might be of benefit to future travellers. Exploring and collecting took their toll of his health and he returned to Kew in 1831 to recuperate and to sort and identify his collections. When his brother Richard, who was colonial botanist and superintendent of the botanic garden at Sydney, was killed by aborigines in 1835, he returned to Australia to fill the vacant post. Three years later he was dead having, in his own words, striven 'to advance, for years, botanic science here from pure love . . . blending the augmentation of our knowledge of the plants of the country with that of its internal geography'.

Notes

[1]Sir William Chambers *Plans . . . of the Gardens and Buildings at Kew* 1763, p. 3.
[2]W Darlington *Memorials of John Bartram and Humphrey Marshall* 1849, p. 232.
[3]F M Bladden *Historical Records of New South Wales*, vol. 3, 1895, p. 202.
[4]*Journal of South African Botany* vol. 25, 1959, p. 185–6.
[5]Banks to Hove 7 January 1787. British Museum (Natural History) BM(NH), Dawson Turner transcripts, vol. 5, p. 122–3.
[6]1787. Ibid, p. 217–25.
[7]25 June 1791. Ibid, vol. 7, p. 218–26.
[8]8 April 1803. British Library Additional MSS 32439, p. 95.
[9]22 February 1791. BM(NH), Dawson Turner transcripts, vol. 5, p. 197–201.
[10]14 September 1795. Ibid, vol. 9, p. 288–91.
[11]7 March 1795, J E B Currey *Reflections on Colony of New South Wales* 1967, p. 8.
[12]27 August 1798. BM(NH), Dawson Turner transcripts, vol. 11, p. 44–5.
[13]29 August 1804, F M Bladden *Historical Records of New South Wales*, vol. 5, p. 460.
[14]18 April 1803. BM(NH), Dawson Turner transcripts, vol. 14, p. 61–8.
[15]6 May 1806. British Library Add. MSS 33981, p. 234–5.
[16]29 May 1814, Royal Botanic Gardens Kew, *Kew Collectors* vol. 1, p. 1–2.
[17]7 June 1814, Ibid, p. 3–4.
[18]1 September 1814, BM(NH), Dawson Turner transcripts, vol. 19, p. 56–63.
[19]April 1818, R B G Kew, *Kew Collectors* vol. 7, p. 32.
[20]*Curtis's Botanical Magazine* plate 2710.

2 From rhododendrons to tropical herbs (1820–1939)

RAY DESMOND

After Sir Joseph Banks's death in 1820 Kew was in serious danger of losing its pre-eminence in the botanical and horticultural world. Allan Cunningham was its only official collector and plants now reached Kew more through chance than planning. When George Aldridge, formerly in charge of the kitchen garden at Kew, was compelled by ill-health to leave Trinidad, he presented Kew with some ferns and orchids. Such offerings, however small, were always gratefully received.

Through the influence of Robert Brown, keeper of the Department of Botany at the British Museum, George Barclay, a young gardener at Kew, was appointed botanical collector in 1835 on H M S *Sulphur* on a survey of the west coast of South and Central America. The botanic results of the voyage, which lasted five years, were disappointing: a few new plants were collected but most were familiar specimens of the coastal vegetation.

When John Armstrong (d. 1847), a gardener at Leigh Park in Surrey, joined the *Alligator* under Captain Brewer in 1837 to help establish a colony at Port Essington on the northwestern coast of Australia, Kew kitted him out as a collector. Armstrong despatched a few plants from Tenerife, Rio de Janeiro, the Cape and Sydney but Captain Brewer frustrated his collecting at Port Essington, insisting that his activities were to be confined to gardening.

In the 1830s the gardens at Kew came perilously close to dismemberment by a thrifty Treasury. They were saved from this fate by the recommendation of an official investigation in 1838 that they should be converted into a national botanical garden. Under pressure from scientists of the day, a reluctant government transferred the gardens to the custody of the Commissioners of Woods and Forests in 1840. In 1841 Sir William Jackson Hooker, professor of botany at Glasgow University, was appointed director of the small botanic garden on the Kew estate. Sir William, blessed with exceptional energy and drive, immediately set about restoring and expanding Kew's reputation as an international botanical and horticultural centre. He dismissed John Armstrong, from whom Kew was receiving no living plants, and, having failed to obtain the Treasury's authority to employ a full-time plant collector, entered into an arrangement with the Duke of Northumberland and the Earl of Derby to share the expenses of collectors.

William Purdie (*c.* 1817–1857), trained at the Royal Botanic Garden in Edinburgh, had been at Kew for less than a year when Sir William selected him in 1843 to collect plants in Jamaica and Colombia for Kew and the Duke of Northumberland. Some of his most successful forays were made in the Sierra Nevada de Santa Marta of Colombia, where despite difficult terrain and torrential rain he made a good haul of orchids. Sir William, seldom satisfied with his collectors' performance, was pleased with 'the number of new, rare and beautiful plants [Purdie]

Sir Joseph Hooker was an accomplished artist as is shown by this beautiful landscape with watercolour tinting. It was done in May 1849 during his Sikkim expedition and features Kinchinjunga (Kangchenjunga 8598 m) looking west-north-west from Simgtam. The original is in the Kew Library.

During Joseph Hooker's Himalayan travels he drew many plants, especially rhododendrons. To save time he painted one or two leaves and flowers, as shown by this one of *R. arboreum* subsp. *cinnamomeum* var. *cinnamomeum* (a). Walter H Fitch, Kew's resident botanical artist, used this as a basis for the fully coloured lithograph (b), which was published in Hooker's *Rhododendrons of Sikkim-Himalaya* as *R. campbelliae* tab. vi (1849).

has been the means of introducing into our Gardens'. The total cost of Purdie's expedition was £1145, of which the Duke of Northumberland paid half, but this amount being more than the Treasury had anticipated, Sir William was instructed to recall Purdie by the end of 1844.

Joseph Burke, who also collected animals for Lord Derby's menagerie at Knowsley Hall, cost Kew only £242. Unable to get to California as planned, Burke explored instead the Snake River area of Idaho and made a bold excursion into Oregon. As very few specimens reached Kew, its impatient and exasperated director ordered Burke to return and was then unwilling to pay his salary. It required a magistrate's arbitration to settle the matter.

Denied a full-time collector, Sir William had to depend on donations. A former pupil at Glasgow, George Gardner, who was superintendent of the botanic garden at Peradeniya in Ceylon (now Sri Lanka), was one of Kew's many benefactors. Through such means Sir William was able to record with some satisfaction in his *Annual Report* for 1850 that between 1843 and 1849 Kew had received 92 cases of living plants, a total that excluded the plants which were just being received from his son, Joseph, in the Sikkim–Himalayas.

In September 1847 Sir William had petitioned the Chief Commissioner of Woods and Forests to be allowed to send a collector to India, whose flora was poorly represented at Kew. He nominated his son, Joseph Dalton Hooker (1817–1911), who had been assistant surgeon and naturalist on H M S *Erebus* during its Antarctic voyage. Treasury approval was obtained and in January 1848 Joseph Hooker stepped ashore at Calcutta, without question the most accomplished of all Kew collectors – a surgeon, botanist, competent artist and able surveyor. His

Gustav Mann was a German who became a Kew gardener in 1859. He went to West Africa on Baikie's Niger Expedition and became the British representative in Fernándo Po, now Bioko. He was the first botanist to explore Cameroon Mountain and sent many dried and living plants and seeds to Kew.

survey work in Sikkim was good enough to form the basis of the Indian Trigonometrical Survey map. He made a number of excursions from Darjeeling but the small kingdom of Sikkim yielded the most exciting discoveries, in particular rhododendrons, covering 'the mountain slopes with a deep green mantle glowing with bells of brilliant colours'. Many of the species he despatched to Kew were successfully raised in a shallow hollow now known as the Rhododendron Dell. After rhododendrons, *Primula sikkimensis* is perhaps one of his most popular introductions.

Charles Barter was a gardener at Kew before becoming a foreman at the Royal Botanic Society in Regent's Park in 1851. Recommended by Sir William Hooker, he was engaged as botanist to the Niger Expedition under W B Baikie. Kew received more than 1300 herbarium specimens of Nigerian plants before his early death in West Africa in 1859. He was replaced by another Kew gardener, Gustav Mann (1836–1916). Unable to join that expedition in the interior, Mann spent three years exploring Bioko Island (Fernando Po), Gabon and Cameroon, forwarding not only herbarium specimens but also at least 25 consignments of living plants and seeds to Kew. Even after he had joined the Indian Forest Service in 1863, he still remembered Kew with boxes of orchids and ferns, mainly from Assam.

When Sir William learnt that the British government was about to present a steam yacht to the Japanese emperor, he obtained permission for Charles Wilford (d. 1893), an assistant in the Kew Herbarium, to be present at the handing-over ceremony, hoping that the propitious occasion might facilitate a botanical expedition by Wilford in Japan. Afterwards Wilford was to join H M S *Actaeon* on its survey of the Manchurian coast (northeastern China). Sir William's letter in October 1858 acknowledging Wilford's first delivery of pressed plants – from Hong Kong – was warm and encouraging. But a little later Wilford was gently reminded by Joseph Hooker, who was by then the deputy director, that he must report his activities regularly. Soon remonstrance turned to exasperation when neither bulbs and seeds nor progress reports reached Kew, although Wilford never neglected to draw upon his expense account. Sir William was about to recall him when the Admiralty appointed him botanist on the *Actaeon* in May 1860. A year later they were similarly disenchanted with his performance and dismissed him.

Wilford's successor, Richard Oldham (1837–64), a Kew gardener barely 23 years old, joined H M S *Actaeon* in mid-1861, now surveying the Japanese coastline. He had three sheets of collecting instructions from Sir William and his earnest advice to 'steer clear' of Wilford, who was still in the East: 'his idling and misconduct has brought him into great difficulties'. Sir William, still remembering Wilford's negligence, was unduly severe with Oldham, who transferred in January 1862 to the *Swallow*, which had taken over the *Actaeon's* surveying duties. Oldham confided to John Smith that Sir William made 'no allowances for the troubles & disappointments I have all along had to encounter'. At the invitation of the British consul on Formosa (now Taiwan),

Oldham visited the island, whose flora was at that time virtually unknown.

When his three-year contract with Kew expired and there were no prospects of further employment, he formally resigned. He had served Kew well, diligently collecting over 1300 herbarium specimens, among them nearly 100 new species. Shortly afterwards he died in China, still bitterly believing that he had been undervalued and underpaid by the Royal Botanic Gardens. He was the last full-time official plant collector employed by Kew.

What were the qualities sought in a potential plant collector? Bowie and Cunningham were informed by Sir Joseph Banks that they had been chosen because they excelled in the 'virtues of honesty, sobriety, diligence, activity, humility and civility'. Bachelors, having no family commitments, were preferred. Sir Joseph was annoyed when George Caley wanted to marry a colonist. 'I did not hire him to beget a family in New South Wales. I fear if he is not more active than is compatible with a married life, I must certainly get rid of him.' Sir Joseph also had a fondness for the Scots. 'So well does the serious turn of a Scottish education fit the mind of Scotsmen to the habits of industry, attention and frugality, that they rarely abandon them at any time of life, and, I may say never while they are young.'[1] And this predilection found confirmation in the most successful collectors: Masson, Menzies and Kerr were all born north of the border and Cunningham was of Scottish extraction.

Some practical knowledge of botany or gardening was essential. Caley was at first rejected as a collector in Australia because as Sir Joseph informed him, 'you are not acquainted with the plants of that country that are already in the gardens. It is important for you to distinguish those that are new in order to send home new plants and leave those already sent home behind, a knowledge indispensably necessary for a botanic collector'.[2] Both Banks and Hooker briefed their collectors very precisely about the plants they were to collect.

Not unreasonably Kew wished to be kept informed of their collectors' progress so letters were encouraged and a journal or diary expected. Collectors risked a reprimand if they strayed for whatever reason from their authorised itinerary. Sir Joseph rebuked Masson in South Africa for such an offence. 'These letters mention your having undertaken 2 long journeys, which surprised me, as your instructions are very absolute on that subject. What I recommend is a fixed residence during the ripening season at any place where plants are abundant.'[3] Sir Joseph also disapproved of Kerr's excursion to the Philippines; he had expected him to remain at Canton. And Oldham earned Sir William Hooker's censure for going to Formosa without his permission. Oldham protested: 'It seems to me to be quite indescribable that a botanical collector who has no special country to explore, should wait for orders from home before he can leave the port at which he may be staying'.[4]

The few reference works collectors were given to help them identify plants were often superficial or out of date. Oldham lamented

Bushes of *Elaeagnus multiflora* were growing at Kew during the 19th century from seeds sent from Japan by either Charles Wilford or Richard Oldham. This drawing was prepared for the *Botanical Magazine* (plate 7341) by Anne Barnard, Sir Joseph Hooker's sister-in-law, and Matilda Smith, who drew the flowers and fruits respectively.

(*opposite page*) One of Augustine Henry's introductions in 1889 was a lily named after him by J G Baker of Kew Herbarium. This original drawing of *Lilium henryi* (plate 7177) is by Matilda Smith (1854–1926), the Kew artist who contributed to *Curtis's Botanical Magazine* for over 40 years.

the absence of an adequate Japanese flora 'to find out the names of plants which I have never before seen'. When Bowie was in Brazil, he wished he had 'the most modern publication respecting the geography and methods of travelling, and manners of the people'.

The difficulties of collecting in a country that was largely unexplored and had poor communications were often fraught with personal danger. Natives could be hostile, as Bowie found in South Africa. Barclay was robbed by bandits high up in the Peruvian Alps and injured by a musket-shot in his thigh; Cunningham was robbed and his life threatened by escaped convicts on Phillip Island, Australia; Masson, a reluctant recruit to the local militia in the West Indies, was taken prisoner by the French; Joseph Hooker suffered a brief imprisonment in Sikkim.

Even after these resolute men got their precious harvest of plants to the nearest port for despatch to Kew their troubles were far from over. Frequently there was no room on any of the ships bound for England. Francis Masson at the Cape was in despair: 'I have many new succulent plants. Some very curious Stapelias, Euphorbias etc. but how I shall get them home, God only knows'. Anton Hove found an East Indiaman leaving India with ample space for his plant cases but the captain had an ex-governor among his passengers and 'could not think of disgracing him with such incumbrances'.

Banks and Hooker gave collectors careful instructions about how to pack their specimens for transportation to England, but the methods employed for packing living plants and seeds were usually fallible, especially with consignments at sea for long periods. A code of accepted practice gradually evolved. When excessive moisture had been removed from orchids and other epiphytes, they were to be packed in a closed box and kept in a cool part of the ship; seeds had to be thoroughly dried; bulbs could be put in boxes of dry sand.

Before the general use of Wardian cases in the 1840s living plants required constant attention. Rainwater rather than water stored in barrels was to be used. Sir Joseph Banks insisted that plants destined for Kew should receive preferential treatment: 'if any other plants except those intended for the King be taken on board, no water shall be issued for them until the King's plants shall have had their full allowance'. Any salt spray deposited on the leaves had to be removed immediately with fresh water. Sir Joseph was convinced that plants stood a much better chance of survival protected in plant cabins or portable glasshouses placed on the ship's deck.

Ships' crews were seldom enthusiastic about the additional task of looking after plants and frequently left them unprotected on deck during storms or neglected to ensure they were watered and ventilated. W T Aiton offered to pay 5 guineas to any seaman prepared to undertake this task; on Banks's recommendation a Chinese gardener occasionally accompanied the plants shipped at Canton; and in 1822 a convict gardener in Australia was promised a free pardon if he succeeded in keeping a shipment of plants in good health.

Francis Masson received £100 a year as a plant collector and nearly a

Bot. Mag.
7177
3.
1.
2.
M Smith del.
Lilium Henryi, Baker.

Portable miniature greenhouses, called Wardian cases, were used to carry cuttings and seedlings on the decks of ships. Originally they were kept airtight but a small hole bored at the end of the case near the top, as in this photograph, increased the specimens' survival rate. They were in use from the late 1830s until airfreight superseded sea transport.

century later the same amount was paid to Kew's last collector, Richard Oldham, apart from a modest expenses allowance. John Livingstone, an East India Company official in China, asserted that William Kerr's 'salary was almost too small for his necessary wants, and he consequently lost respect and consideration in the eyes even of the Chinese assistants, whom he was obliged to employ'.[5] Sir William Hooker always exhorted his collectors to practise 'prudence, economy and judgement' in their expenditure.

Before Bowie and Cunningham left for Brazil, Sir Joseph Banks blandly assured them that if they were frugal 'your reward is within your reach. You will be able to return to your native country before the afternoon of life has closed with a fair prospect of enjoying the evening in ease, comfort and respectability'.[6] Some did return safely and in good health, a few found advancement in colonial botanical gardens but others were less fortunate. Masson died in Canada, Nelson in the Pacific and Barter in West Africa; dysentery killed Good and Oldham; some believed that Kerr became an opium addict, which may have hastened his death in Ceylon. These men accepted the danger as well as the glamour of plant collecting and with few exceptions served Kew loyally and well.

Kew-sponsored private collectors

After Oldham's death in 1864, the acquisition of plants for Kew depended upon the generosity of individuals and exchanges with other botanical gardens and large nurseries, such as Messrs Veitch of Chelsea.

In March 1885 Kew received some seeds and a request for botanical advice from a young customs official in China. Augustine Henry (1857–1930) followed up his offer to send Chinese plants to Kew with the first of many consignments of herbarium specimens in December of that year. By 1900 Kew had received from him some 16,000 specimens, including over 20 new genera and 500 species. A favourite with British gardeners is *Lilium henryi*, which Henry discovered growing in limestone cliffs in Hupeh Province.

A Kew gardener, William Lunt (1871–1904), who accompanied Theodore Bent's expedition to Hadramaut in South Arabia (now Southern Yemen) in 1893, came home the following year with 150 species for Kew.

Techniques for shipping plants and seeds were still in some respects no more sophisticated than those adopted by earlier collectors. The *Kew Bulletin* for 1914 recommended that seeds should be placed in paper or canvas bags; bulbs, rhizomes and tubes were to be dried and packed in boxes of shavings or any other light material; woody-stemmed plants, capable of withstanding drought for several weeks, could be packed in ventilated boxes. Wardian cases were reserved for plants that could not be sent by any other means. For many years the Wardian case was kept absolutely airtight but experience showed that the survival rate of plants improved if small holes were bored at each end of the case near the top.

Augustine Henry (1857–1930) was not a full-time Kew collector but he sent back many consignments of dried plants from China, and some seeds and bulbs.

The obvious abilities of John Hutchinson (1884–1972) gained him an early transfer from the gardens at Kew to the Herbarium, where he eventually took charge of the African Section. A compulsive traveller, he never missed any opportunity of seeing the world's flora. He and a Kew colleague went botanising in the Canary Islands in 1913 and brought back over 600 specimens for the Herbarium. A grant from the Empire Marketing Board financed a nine-month trip to South Africa in 1928–9. He found travelling in the comparative comfort of a car had its drawbacks. 'Whilst a car may be regarded in many respects as a very great asset to the botanical collector, it depends largely on the willingness of the driver, even though he be a botanist, to stop on every occasion demanded by the untrammelled eyes of the passengers.'[7] He returned to South Africa in 1930 to join General Smuts on a botanical expedition to Northern Rhodesia (now Zambia) which added some 2000 specimens to the Kew collections, including a remarkable new species in the daisy family (Compositae), *Pteronia smutsii*.

When Kew botanists went on holiday they often came back with some floral offering for the Herbarium. The Spanish collection was enriched by N Y Sandwith's frequent visits to that country. Noel Sandwith (1901–65), who was in charge of the American Section, made two important collecting expeditions to British Guiana (now Guyana), the first in 1929 as a member of an Oxford University expedition and the second in 1937 as botanist to an Imperial College of Science entomological expedition. Knowing first-hand the value of collecting experience, he bequeathed a sum of money to enable Kew botanists under the age of 40 'to make a collection of dried plant specimens for the Kew Herbarium in any tropical country'.

An industrious six weeks' collecting in central and southern Spain in 1924 by the Revd E Ellman and Charles E Hubbard (1900–80) yielded more than 1000 specimens for Kew. When Hubbard was seconded to the Brisbane Herbarium in 1930 to revise their grass collection, he took the opportunity to travel widely in Australia, sometimes in regions explored a century earlier by Allan Cunningham. It has been estimated that, with many duplicates, he collected some 100,000 specimens of grasses, a family on which he subsequently became a world authority.

Times and methods were changing, however, for in 1930 Edgar Milne-Redhead joined the ground team of an aerial survey in Northern Rhodesia. He returned there in 1937–8 and around Mwinilunga he made an outstanding collection of new and little-known plant species. His activities, and those of John Hutchinson, heralded increasing Kew-based plant hunting in tropical Africa which, as we shall see, had to wait until after the Second World War for its full blossoming.

Notes

[1]14 September 1814 British Museum (Natural History) BM(NH), Dawson Turner transcripts, vol. 9, p. 56–63.
[2]27 August 1798. Banks papers in Mitchell Library, Sydney.
[3]3 June 1787. BM(NH), Dawson Turner transcripts, vol. 5, p. 173–4.
[4]8 June 1864, Royal Botanic Gardens (R B G) Kew, *Kew Collectors* vol. 9, p. 44–6.
[5]*Transactions of Horticultural Society of London* vol. 3, 1822, p. 424.
[6]18 September 1814 R B G Kew, *Kew Collectors: Bowie & Cunningham*, p. 15.
[7]*Kew Bulletin* 1929, p. 277.

A modern satellite view of Africa.
Photo: METEOSAT image supplied by the European Space Agency

PART II

RECENT KEW EXPEDITIONS AROUND THE WORLD

3 *Tropical Africa continues to surprise*

In spite of being relatively close to Europe and on the sea-route to India, tropical Africa remained the Dark Continent and little was known of its flora and fauna for many centuries. In the 1860s, however, Kew launched its multi-volume *Flora of Tropical Africa* (1862–1932), although exploration had hardly started by that time. In the present century Kew has sponsored the *Flora of West Tropical Africa* (ed. 1 1927–36, ed. 2 1954–72) the *Flora of Tropical East Africa* (1952–) and *Flora Zambesiaca* (1961–). The last two are still continuing with associated fieldwork. In fact, it is true to say that during the middle decades of the 20th century tropical Africa was the main thrust of taxonomic reasearch at the Kew Herbarium. Staff were specially recruited for the purpose.

Three very different accounts of fieldwork in tropical Africa follow, showing how varied Kew expeditions can be in their objectives, as well as in their unexpected and exciting experiences.

Two treks in West Africa

F NIGEL HEPPER

West Africa had the unenviable reputation of being the 'white man's grave', yet it was in the 19th century from coastal forts, riverside plantation-settlements and forest mission-stations that the first dried specimens found their way to Kew Herbarium. Sir William Hooker edited the *Niger Flora* (1849), which was based on the plants collected by Theodor Vogel, who was sent by Kew on the ill-fated 1841 Niger Expedition. By the 1920s, when John Hutchinson and J M Dalziel were writing the first edition of the *Flora of West Tropical Africa*, British and French foresters and agriculturists were training local staff to study the flora. Following the Second World War there was an urgent need for a complete revision of that work and in 1951 R W J Keay of the Nigerian

Map 2 Western Africa with places mentioned in 'Two treks in West Africa', ch. 3.

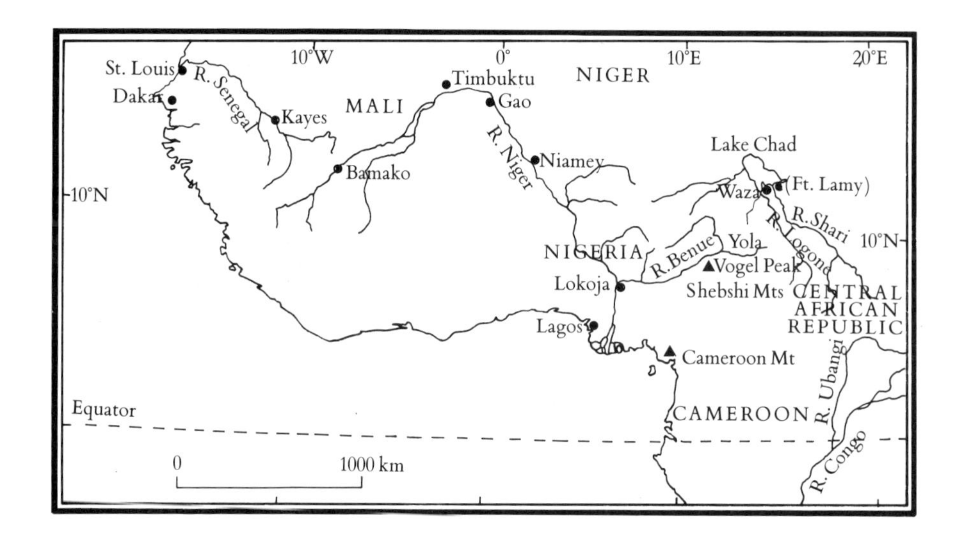

(*opposite page*) The extensive savanna zone of tropical Africa is dominated by grasses and herbs, with small trees well spaced out. In the savanna of northern Nigeria bordering Cameroon where this expedition trekked, an unusually bright red form of *Vernonia nigritiana* (Compositae) was accompanied by the slender white-flowered *Lippia rugosa* (Verbenaceae).
Photo: F N Hepper

Forest Service came to Kew for that purpose. Having completed the first volume, he returned to West Africa in 1957 and I continued to edit the last two volumes until 1972.

At first, field experience was necessary, so in 1957 Kew sent me to the former German colony of Cameroon, then administered as a British mandate under Nigeria. The Berlin Herbarium containing millions of specimens had been bombed and I was hoping not only to go to the same areas visited by German expeditions and re-collect lost plants, but to increase our botanical knowledge of a little-known region at the eastern edge of the *Flora* area.

Trekking with carriers

After 13 days at sea from Liverpool I arrived at Lagos harbour. Some 1400 km (850 miles) northeast of Lagos lies the hot and dusty town of

On the Shebshi Mountains and on Vogel Peak in Cameroon old-fashioned trekking was the only and best way of getting about. A tent and tin bath containing a basket full of clothes, as well as numerous plant presses and boxes of food and equipment, were transported by the local carriers for several months in 1957–58.
Photo: B O Daramola

Yola, where I recruited porters (or 'carriers' in West African parlance) to bear my equipment. A technician from the Ibadan Forest Herbarium, Benjamin Daramola, had joined me for the expedition and we set out by lorry across the savanna plain dotted with *Parkia* and shea-butter trees, towards the distant Shebshi Mountains, with Vogel Peak in their midst. At the end of the motorable road the carriers took to their feet and I to a horse. The following day the escarpment loomed steadily nearer, topped by clouds which indicated that the rainy season was still not spent.

I was anxious to collect as many plants as possible in the lowlands before the heat of the dry season scorched them. The sorghum and maize fields yielded interesting annuals (*Kohautia, Oldenlandia, Striga*), and the rice swamps aquatic herbs such as *Utricularia, Xyris, Drosera* and *Eriocaulon*. In the drier savanna there were numerous leguminous herbs in the genera *Crotalaria, Indigofera* and *Tephrosia*. Overhead, crowned cranes honked as they flew past, Egyptian vultures circled higher and sacred ibis dabbled in the river where natural woody vegetation persisted along the banks.

After a week based at a thatched mudhut we set off for the hills. I had assembled some 20 carriers to transport enough collecting equipment, tinned and dried food, tents and bedding for a month's trek. Wooden presses, drying paper and flimsies (for separating the drying specimens) were heavy and bulky; jars for delicate specimens and spirit for pickling them presented their own transport problems; valuable notebooks and bags of coins to pay the villagers were secreted in unidentifiable boxes. The carriers were also paid in coins – 6 pennies a day 'chop money', with the balance of about 2 shillings retained until the end of the trek.

At first light the head loads were equalised by the man in charge and suddenly the carriers were off, padding barefoot for the escarpment, encouraged by the senior man's sheep-horn, on which he blew repetitive mournful notes. The speed slackened as the gradient increased and the sun rose. As gleaming bodies reached the crest and rested, we noted the changed flora of the higher altitude in thickets and rocky outcrops, grazed but undisturbed by agriculture – shrubby *Erythrina, Hymenocardia, Hypericum* and *Dalbergia* with the occasional wild African banana, *Ensete gilletii*, and rock-inhabiting *Aeolanthus* and *Cyanotis*.

Southwards rose the great massif with several summits, including Vogel Peak itself. A beautiful valley lay between the mountains and a plateau to the east – it was to this valley that we had to drop for supplies for the men and, as we descended, the sound of the headman's whistle far above us signalled trouble, so a guide was sent back to investigate. One carrier was too ill to continue. All the others were strung out on the mountain path and my yellow boxes could be seen bobbing along. They reached the first cluster of huts, where the women were grinding sorghum by rubbing the grain on a stone outcrop.

Several tiny villages nestled among the patches of tall sorghum (Guinea corn) and we based ourselves at one called Dakemi. There was a shortage of this ripe corn to supply the carriers' needs so it was fortunate that a cow was accidentally killed as it fell from the precipitous path from the plateau and its meat was sold in the village. Large herds were gathering and descending to the valley, seeking grazing along the rivers further south as the dry season reduced the pasture on the hills. Considerable damage was done by the cows en route and one villager implored me to obtain compensation not only for his crops consumed but for his wife, who had been abducted by the nomads!

The grassy mountainsides were peppered with bushes of *Combretum, Terminalia* and *Uapaca*, while the herbaceous flora was rich and varied, including *Polygala* and a *Justicia* which proved to be a new species (*J. hepperi*). Head-high *Hyparrhenia* grass grew in the moister areas, smaller *Setaria* on the slopes and dwarf *Ctenium* in the dry rocky places. Already, however, grass fires were being started to promote new growth for the remaining cattle, and there was no time to lose. I continued collecting wherever I went. A general collection of everything in flower and fruit is valuable from this type of remote area, whereas only special, selective gatherings are wanted at Kew from better-known regions.

The route to the peaks was only resolved by enquiring at each village since the existing map was at far too small a scale to be useful. By trekking around the massif and viewing the tops from the plateau, I was able to see what was possible. The plateau itself yielded another plant novelty: in spite of having the appearance of a cauliflower, it was a *Combretum* which was given the name of *C. brassiciforme*. It was also there that we met lion trouble! Nine cows had been killed and the farmer was fearful and distracted. His small son was ill, yet he was wandering around with a fever which I diagnosed as pneumonia and I prescribed rest and gave aspirins, for which we received a gourd-full of milk.

Somehow a letter found its way to me simply addressed to 'F N Hepper Esq Botanist-at-Large'. It was from the district officer, who was keeping an eye on my welfare – just as well when I nearly had a murder on my hands. One night the carriers made merry with the local brew, tempers flared and a knife was drawn. The whole affair was safely defused after I called a midnight assembly and gave a stern warning to the men.

Vogel Peak, rising as a hump in the central portion of the massif, was set back from the main valley and it could only be approached from a side valley. On reaching a hamlet there, it was very difficult to find a villager willing to act as guide since the last one to venture to the summit had fallen ill and died. I took some of the carriers with plant presses and tents, the ascent led through tall riverine forest with a tangle of hauser-like lianes. The rocks beside a high waterfall were red with a species of Podostemaceae, a small tropical family highly adapted to fast-flowing rivers with a variable level according to season – the lichen-like plants encrust the rocks where small flowers appear in succession at water level.

On reaching the upland meadows harnessed antelopes bounded away. Wild *Gladiolus, Urginea* and other bulbs were in flower and a dwarf *Thesium* and rock plants clustered among the scattered *Protea* and *Psorospermum* bushes. The carriers soon erected my tent, and the shade of an old fig tree served as a kitchen with a camp fire and baking oven composed of old kerosene tins.

Clouds swept over the tops some 2000 m (5600 ft) high and we caught distant views across the plateau as we headed for each of the summits in turn to determine which was Vogel Peak. Troops of baboons accounted for the unexpected paths between them. An outlying clump of trees by a steam yielded the first West African record of a hitherto East African *Crotalaria* – something that occurred repeatedly as the highland collections increased. Nearby a large *Podocarpus milanjianus* tree confirmed the upland nature of the vegetation and further north than previously noted.

Having climbed Vogel Peak, it was necessary to return to the hot main valley and rejoin the carriers with the main part of our loads. Collecting continued along the escarpment for a further fortnight, with camps being established at several places. By then the collections had assumed considerable proportions with some 450 gatherings, mostly in quadruplicate (i.e. 1800 dried specimens, each in a separate flimsy folder), plus spirit jars and bags of bulky dried fruit.

As the dry season set in we continued the expedition further south for two more months, losing the carriers for a worrying five days, walking across the Mambila Plateau to the moister Bamenda Highlands and leaving Yola far behind. It was to be some 11 years before I revisited that town in very different circumstances.

Collecting by hovercraft

When the director of Kew, Sir George Taylor, heard about the proposed Trans-African Hovercraft Expedition organised by the Royal Geographic Society, he arranged for me to join it at Dakar, Senegal, in September 1969. Here was a splendid opportunity to visit the mainly francophone countries covered by the *Flora of West Tropical Africa*. The expedition's route from Senegal followed the Niger through Mali, Niger, Nigeria and Cameroon, continuing to Lake Chad and the Central African Republic, and it would be possible to collect material en route in areas seldom visited by Kew botanists.

Two SRN6 hovercrafts were expected to take part, but in the event only one was hired owing to lack of funds. With too many expedition members to travel in one vehicle, other arrangements had to be made for each leg of the journey. This proved to be a blessing for the scientists and geographers including Professor R Harrison Church and A T Grove of Cambridge University, as the more leisurely pace in a UNESCO survey pontoon enabled us to make frequent stops during the 600 km (373 mile) trip up the River Senegal to sample the vegetation and to study the irrigation and agricultural schemes beside the river. African villages and delapidated ex-French villas were

Collecting plants from a suspended hovercraft was a novel and rather hazardous technique developed along the West African rivers during the Trans-African Hovercraft Expedition in 1969.

scattered along the banks, Mauritania one side and Senegal on the other, and we received esctatic welcomes from paddling canoeists. The local inhabitants were puzzled by the non-appearance of the heralded hovercraft, which eventually overtook us and at Kayes, the limit of navagation, had to be dismantled for transit by rail to Bamako on the River Niger.

From Kayes the expedition members travelled in a primitive waggon on the same freight train, which nearly met with disaster as portions of the SRN6 came adrift. Only prompt action stopped the locomotive on a bridge where we could have been derailed into the gorge – but it afforded me a chance to botanise in an unexpected locality while the errant load was secured. The remainder of the 19-hour journey was enlivened by a spectacular night-time thunderstorm among baobab trees and flat-topped hills.

While the hovercraft was being reassembled, we continued down the River Niger in a three-deck ferry for Timbuktu in Mali and Gao in Niger Republic. Long stops enabled further collecting, even in the dry Sahel zone as far from the river as Bandiagara, where the Dogon people inhabit a harsh rocky environment. Some days at Gao were profitably occupied among the aquatic vegetation of the extensive swamps adjacent to the Sahara sand-dunes.

Joining the noisy hovercraft, we sped down the Niger through several types of vegetation from the Sahara desert, and Sahel and Guinea savannas to almost the rainforest zone. Instead of continuing to the Atlantic Ocean via the delta, our route took us from Lokoja up the River Benue north-eastwards towards Lake Chad. The massive dry-season sandbanks and rapids were covered with ease by the hovercraft, much to the astonishment of the local people and the hippos. In a few days I was once again at Yola, where the lamido (emir) recalled my previous expedition – and now I was able to invite him into a vehicle that had not then been invented.

Garoua, in Cameroon, is the end of navigation on the River Benue and once again the SRN6 had to be dismantled; this time for lorry transport to the River Logone, which flows into Lake Chad. Meanwhile, we scientists travelled by road through the Mandara mountains, where fantastic denuded volcanic cones rise high above the grassy slopes rich in montane plants, such as the orange-flowered asclepiad *Stathostelma pedunculatum*, and peppered with cactus-like *Euphorbia deightonii*. At lower altitudes nearer Fort Lamy (now N'Djamena) the game reserve of Waza supports taller savanna vegetation, elephants and giraffes.

On reaching Fort Lamy I cooperated with a French research botanist in collecting excursions on the shore of Lake Chad. Later, I flew in a missionary floatplane over this huge shallow lake. At that time the lake covered an area about the size of Wales, and numerous linear islands formed from flooded sand-dunes were fringed by dark papyrus swamps. I stayed on one of these islands and sampled the dry interior of *Acacia, Leptadenia* and *Calotropis* bushes, as well as the marginal reed beds of *Phragmites* and papyrus *Cyperus papyrus*. When the hovercraft eventually reached the lake, following repairs after an accident, we were able to investigate the lush vegetation growing on natural papyrus mats which had drifted away from the islands, and reach areas too shallow for power boats.

How different was this expedition using sophisticated technology from the traditional foot safari to Vogel Peak. Many years later, during the mid-1980s, my involvement with Cameroon continued in yet

After a missionary aeroplane had deposited Nigel Hepper on a sandbank in Lake Chad, he used a dugout canoe and papyrus skiff to study the waterside plants fringing the islands.
Photo: F N Hepper

another way when I initiated a rainforest conservation project sponsored by the Overseas Development Administration in conjuction with Kew and the Cameroon government. Based at Limbe (formerly Victoria) Botanic Garden, the project is on the volcanic slopes of Mount Cameroon, a unique site, extending from sea level to altitudinal limit at 2900 m (9500 ft) with grassland to 4070 m (13,350 ft). It is setting an example of international cooperation which has great potential for the conservation of genetic resources in Africa.

When it rained in East Africa

R M POLHILL

Extracts from correspondence and journal of Fourth Colonial Office Expedition to East Africa

From the Secretary of State for the Colonies

To the Administrator, East African Herbarium *17 April, 1961*

Mr R M Polhill – East African Herbarium

The Director of the Royal Botanic Gardens, Kew, is of the opinion that there would be advantage in Mr Polhill carrying out a collecting expedition in East Africa before he comes to England to take up duty in the Herbarium at Kew. Dr Taylor [the director] has written as follows:-

'I am advised that priority should be given for a good collector to spend some time on the Great North Road in Tanganyika between Kondoa and Sao Highlands during the rainy season, which in that area is mainly from December to April. . . . In October he should be asked to go to Garissa in the Northern Frontier District of Kenya and collect along the Tana River and in the adjacent open country during the small rains – November and December. He should return to Nairobi in December and have Christmas at home. He should then go to Kondoa, reaching there early in January, working up and down the road and staying at one or other of the townships. We would like him to visit Mufindi and make a good collection in the tea growing area along the top of the escarpment. We suggest a

Prolonged torrential rain flooded the expedition campsite at Kurawa, Kenya, in September 1961.
Photo: R M Polhill

visit be made to the point where Goetze [a German botanical explorer] crossed what is now the Great North Road (at the same season) so as to try and obtain some of the rare species Goetze collected thereabouts at the end of the last century. The safari should end about the middle of May and Mr Polhill might be expected to start work at Kew in June 1962.'

East African Herbarium
4th July 1961

E Milne-Redhead Esq.,
Head of the African Section,
Royal Botanic Gardens, Kew

Dear Mr Milne-Redhead,

Thank you for sending off the miscellaneous equipment from the Kew stores.

As you may well imagine the Colonial Office jibbed at giving me a new vehicle, but through the High Commission I have found a second-hand 2-ton diesel Ford truck, with 30,000 miles on the clock, recently overhauled and with good tyres. I am satisfied it will do perfectly well. They have had a new body with a canvas top put on and added a few simple fittings – a hold for three 40 gallon drums (two water, one fuel), a bin for 50 [plant drying] presses and 70 cu. ft. of flanking cupboard space towards the rear, the tops forming a table on one side and a sleeping bunk on the other. The relatively small additional cost will I hope reduce the fuss of making and breaking camp. The lorry will cost only £250 (such are present prices and in addition the firm receiving the vehicle in part exchange is willing to resell it without profit) and the body and fittings a further £200.

I have been put in touch with the Desert Locust Survey and they have offered me the facilities of their base camp at Garissa while I am in that area. I only hope we get decent rains at the normal time.

Yours sincerely,

R Polhill

[JOURNAL]

15 September 1961

Left Nairobi 7.30 a.m., reaching Mtito Andei, 149 miles, at 4.15, collecting 12 numbers [i.e. different specimens], pitched short camp [i.e. without work tent]. Vehicle loaded takes small corrugations of road splendidly, most comfortable, engine temp. consistently low, no dust in rear, cool in cab. Two assistants first class – invaluable attribute of taking pride in work well done. Field notes typed, but very tired by end. Sleeping accommodation all round satisfactory. Everywhere appallingly dry.

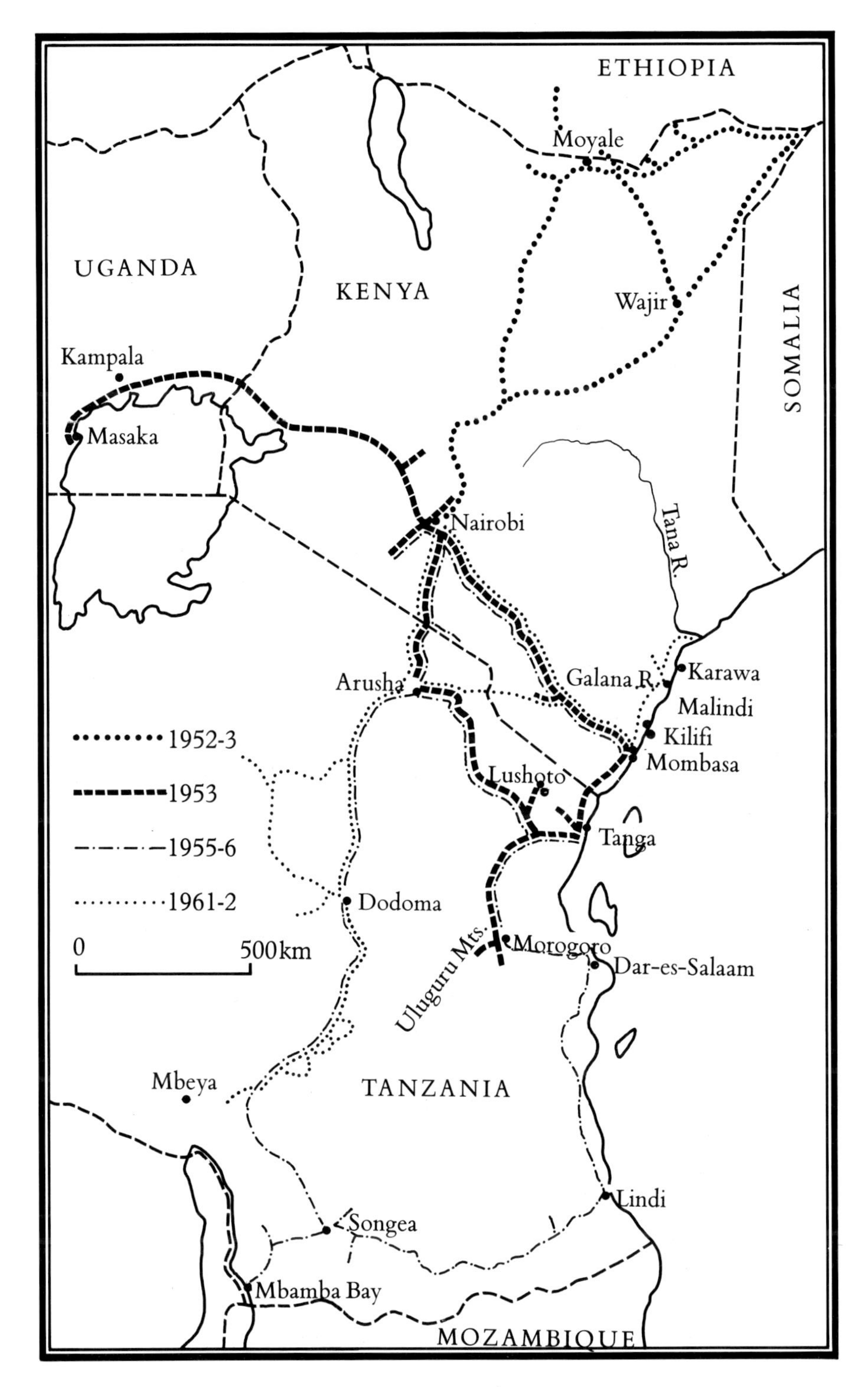

Map 3 Soon after the end of the Second World War the Colonial Office sponsored the *Flora of Tropical East Africa*, with botanists based at Kew. In 1949 the opportunity arose for one of the senior botanists, J B Gillett, to join the Kenya–Ethiopia Boundary Commission. Intensive collecting from March 1952 to February 1953 resulted in 1838 herbarium collections, including numerous new species and first records of genera and species not hitherto recorded from Kenya.

The success of the expedition persuaded the Colonial Office to mount a second expedition in 1953. R B Drummond and J H Hemsley made an extensive tour, spending the greatest part of their time in the mountains of eastern Tanganyika (now Tanzania) and along the southern coast of Kenya. In the nine months from January 1953 they made 3791 collections and revisited many localities frequented by German collectors before the First World War.

In 1955 E Milne-Redhead, editor of the *Flora*, and P Taylor travelled via Nairobi down to southern Tanganyika, making an outstanding collection, principally in the Songea District. The specimens, totalling 3407, were prepared with meticulous care and remain by far the most important collections from this remote area. At the end of their ten months they returned rapidly up the Great North Road to Kenya, gaining the impression that this cross-country route would repay more intensive collecting.

Hence the proposal for the 1961–2 expedition whose initial stages are described below. Between August 1961 and May 1962, 1736 collections were made, bringing the total number derived from the Colonial Office expeditions to 10,772 deposited at Kew, with duplicates distributed to the East African Herbarium in Nairobi and a number of other herbaria in Europe and Africa.

19 September 1961
Travelled until early afternoon into dry country between Malindi and Garsen and turned off main road to Karawa, 30 miles south of Garsen, and got thoroughly stuck in black clay soils – after 2–3 hours with little success using jack, chains and wood under wheel, walked to the livestock holding ground a mile down the road and borrowed their tractor to pull us out and set up camp in their compound.

25 September 1961
Very good morning collecting . . . Started to rain heavily at about 11 o'clock and only collected a few more between squalls, returning to camp about 4 o'clock. Rain had prevented satisfactory drying of drying papers and much of little remaining charcoal very wet.

Standing water near camp attracts zebra and water buck and at night the odd lion which make themselves felt by considerable noise – Yohana, not used to safari work, scared out of his wits, bundles Samwell and himself into the lorry until nearly dawn and I receive a long and amusing harangue on their fierceness, hunger and proximity next morning. Now arranged for them to sleep in the lorry.

26 September 1961
Awoke to continuing rain, whole of the vlei [shallow valley] below the compound flooded and water still rising. Rain gauge recorded 8.65 inches for the 24 hours. No drying paper usable and still raining. Typed field notes first half of the morning by which time flood water up to camp site, drowning fire and channels of the charcoal burners. Still pouring with rain – everything in tent soaking. Moved all into back of lorry.

Kindly offered use of part of wireless hut – tin shack on higher ground. On investigation approach road much too deeply under water to take lorry. Stood about until 4 o'clock, shivering, but no let up of rain and flood water up to axles of lorry. Gave up for the day – piled into lorry, reshuffled all equipment, changed, fed communally over a primus in rear of lorry, early to bed – still raining.

27 September 1961
Still raining – gauge recorded a further 7½ inches – 17 inches in 2 days! Cleared a little about 9 o'clock and labour gang cheerfully moved all necessary equipment ¼ mile to tin hut. Camp site completely under water and sheets of water in all directions – now apparent as overflow from the Tana [River]. Transmitter now next door and overhearing, rain widespread, people stuck everywhere, even in land-rovers, main bridge across Sabaki River at Malindi

under water and impassable. Drying paper stacked in chests but mostly not dry by evening – some specimens beginning to suffer – made as good a change as possible.

28 September 1961
Cleared up in morning and managed to salvage and dry tents. Water up exhaust of lorry. Drying paper all out in sun and caught by sudden squall and all soaked again. Charcoal now nearly finished. Beginning to get really worried about lorry – almost all labour left for Malindi for pay and to collect rations, walking all 50 miles and presumably ferried across Sabaki River. Hear on radio that bridge now swept away and messages all over trying to get people out of the Tana River District. Left here are 6 men and a Fordson tractor.

No plant collecting – nearest dry ground at least one mile away and will need full day for useful work. Preservation of material already collected more important.

29 September 1961
Rained and water rose considerably overnight. Much too deep between here and lorry to use tractor. All available men tried to push it, but could budge it no more than a few inches, water now covering number plates and entering tool box. Decided wireless hut likely to be flooded so packed up everything and carried to main store about 40 yards away. The major operation of ferrying essential camp equipment to much higher ground about 1 mile away – seemed much more! – most of the work was literally on our three heads. All very tired by the time camp set up in the evening.

30 September 1961
Water even higher and now nearly into cab of the lorry.

1 October 1961
Water rose considerably last night and now almost into body of the lorry . . . Removed all stores, occasional squalls and water choppy and at last load tried to put on too much, boxes slipped off and in rescuing them tyre used as a raft caught in wind and rapidly blown away.

2 October 1961
The camp site on the hill is now somewhat threatened, the water is still rising, if perhaps a little more slowly and the end of the island where we are forms a promontory at not quite the highest point. There is a strong flow of water behind and if this should top the brow will flood down on the camp site.

The expedition lorry with water up to its axles and bodywork, carefully making its way along the track from Kurawa to Malindi, 31 October 1961.
Photo: R M Polhill

3 October 1961
Water rose much less last night and perhaps little further to worry about. Plane came over at 9 o'clock and dropped tentage and food and a battery for the radio, for the compound staff. Great excitement – tent soon up. Pottered around and collected 6 plants – everyone rather restless.

4 October 1961
Malindi did not hear us on the wireless yesterday and sent a helicopter to enquire after us at breakfast time. Assured them we were all right and confirmed by radio at 8 o'clock.

Indicated that we should be prepared to be evacuated, so spent morning repacking all water-damaged articles in tin boxes and sorted out things to be taken out. At 12 o'clock, however, told that wireless and two of the compound staff to be left behind and the rest collected by boat – dhow up the coast and then a punt inland.

Did I want to leave? Decided I might as well stay another couple of weeks and see how the lorry salvage position would be – assistants disappointed. Spent rest of the day writing letters. Rained in the night and tent collapsed, which was very unpleasant.

E Milne-Redhead Esq.,
Royal Botanic Gardens, Kew

Kurawa
Thursday 12 October

Dear Mr Milne-Redhead,

I am glad to say things have improved here greatly since my last letter, which was delayed in posting. The boat didn't make it, but the Livestock Officer got through by tractor yesterday. If it dries up a bit more in the next few days, he hopes to send out a trailer which can take mail on its return.

The water has fallen very nearly as fast as it came up, the Tana is now back in its bed and the water covers the lorry only up to the wheel hubs. We spent some time today going over the bodywork with wood preservative and applying superficially oil and grease where we could. The Veterinary Department hope to send out a Ministry of Works mechanic to go over all their own machinery and have kindly included my vehicle on the list. The engine will have to be stripped and the rest gone over carefully, but otherwise I can see no obvious damage.

We have collected the island pretty exhaustively and the water is now low enough to wade on to the mainland, but the bush is all much alike and it is becoming difficult to find species we have not yet collected.

I shall keep you informed as far as the postal system allows.

Yours sincerely

R.M. Polhill

Royal Botanic Gardens, Kew
17th October 1961

Dear Polhill

I got your letter telling me of the sad story yesterday, and to-day your more cheerful one of 12th October. I am so sorry to hear of your troubles. The conditions must have been quite fantastic and I am sure no one could possibly hold you in any way responsible for what has happened. I am glad you are none the worse yourself and that you have saved your equipment. It is good that the veterinary department are so helpful and I hope the Ministry of Works mechanic will be able to get your truck back to life.

Don't worry! Having collected more obvious things, you are doubtless finding some of the less conspicuous items (but none the less interesting) which you would otherwise have missed.

With kind regards,

Yours sincerely, in haste

E. MILNE-REDHEAD

At last the new ferry over the River Sabaki, Kenya, started to operate on 6 December 1961 and transported the expedition lorry. Fritz Bauer, the Livestock Officer for the Veterinary Department at Malindi is in the centre of the group – he is typical of the kind of person who is an invaluable help at times of difficulty during Kew expeditions.
Photo: R M Polhill

31 October 1961
Finished packing up lorry, stacking everything off the floor-boards in the back, preparing for deep water. Left half-past eight for Malindi accompanied by tractor and trailer [**19**]. Lorry sailed through all the deep water, up to our shins in the cab at places, sandy soil fortunately forms no mud but did get stuck once in sand drift, stalling engine with battery too flat to restart and at one bridge rebuilt with red soil gave on side and swerving too much the other way got caught in very soft soil in which I had no control over steering. Into deep ditch and lorry at hideous angle, but hauled out by tractor and people pushing. Reached the Sabaki River mid-day only to find that the river too high and fast for the Royal Engineers' home-made ferry to work. Set up camp and then crossed the river by canoe and spent the night with the Brauers. [Fritz Brauer, the Livestock Officer for the Veterinary Department at Malindi, for some time just a reassuring voice on the radio, had been back and fourth to Kurawa several times in that last three weeks, befriended me and got my vehicle on the road again. He and his wife Jill were most generous with their kindness and hospitality, and organised a number of local excursions for me.]

17 November 1961
Went shopping in Malindi and had lunch with Fritz and Jill. Immediately afterwards heard from the District Officer that the cableway over the Sabaki had gone! Went down and found whole bank collapsing right up to the trestle supporting the cable on the Malindi side. Overseer agreed to make one more trip and ferried me across. Fritz phoning for permission for me to take a land-rover and trailer up to Marafa. Sat about in nervous state most of rest of the day.

18 November 1961
Packed but permission to use land-rover did not come through until noon. Drove straight to Marafa to avoid rain. Road OK but very narrow and cut up, only passable to land-rover. Idyllic spot in *Brachystegia* woodland, rest house on a bluff overlooking the river with a view of bush to the distant horizon, cool and very clean after Sabaki. Model village with a chief straight out of Bafut Beagles [a book on collecting in Cameroon by Gerald Durrell], who insisted on showing me around and collected beautiful bore-hole water three-quarters of mile away. How wonderful to be able to leave things out on tables etc., leave bedding unrolled and not continually repacking. Started collecting.

19 November 1961
Splendid day collecting in the damper forest down to the river – 20 numbers.

20 November 1961
Another good collecting day in forest along road to bore-hole and in the afternoon up a dry river bed into the incredible erosion scarps found on the Magarini sands hereabouts.

21 November 1961
Long walk but not very satisfactory collecting – 9 numbers in all. Meditations on my 24th birthday.

East African Herbarium
7 December 1961

Dear Roger

Have just returned from 3 days holiday to find your letter in the pile. I weep for you – all the time you have had to waste. Do stop at Voi and particularly look for a small *Moringa* a yard or so tall.

I was annoyed about the land-rover business, but when you rang the road to Mombasa was quite closed and the Namanga road likewise, also two EAAFRO [East Africa Agriculture and Forestry Research Organisation] land-rovers were bogged down at Machakos and elsewhere.

. . . Very glad you have managed to find more helpful folk down there and have been able to travel afield a little. The two parcels arrived safely and were checked.

. . . Very glad to hear that a ferry will soon be working and that this letter may arrive too late. Let us hope that the rest of your expedition will not be dogged by bad luck. I hope these setbacks will have not been too expensive, but suspect they may have been. Best wishes from us all.

Yours

BERNARD VERDCOURT

A new ferry was constructed and brought the lorry across the Sabaki on its first day, 6 December. Our subsequent expedition was successful, returning to Nairobi for Christmas and working in Tanganyika (Tanzania) until May 1962. The abiding recollections are of profitable collecting, good companions and outstanding kindness and hospitality wherever we went, but that season the rains never seemed to stop. When they did – in May – the central plateau of Tanganyika was a sheet of brilliantly coloured flowers as far as the eye could see.

After the rainy season, vegetation quickly comes into flower before the scorching sun dries it up for another year. Here the Central Plateau of Tanzania in May 1962 was a beautiful sight after the prolific rains earlier in the year, and members of the expedition made collections of plants seldom seen except in wet seasons and rarely collected for Kew.
Photo: R M Polhill

The Malawi connection

R K BRUMMITT

The travel brochures these days proclaim Malawi to be the 'Warm Heart of Africa', a claim amply justified by the cordial reception its friendly people give to visitors. To the botanist there is the additional attraction of a rich and varied flora, often set in magnificent scenery of high plateaux or lake shore. For Malawi cradles the southern end of the great Rift Valley system of eastern Africa between its mountain massifs, and within 40 km (25 miles) on a map one can pass from the heat and humidity of the lower Shire Valley at a mere 30 m (100 ft)

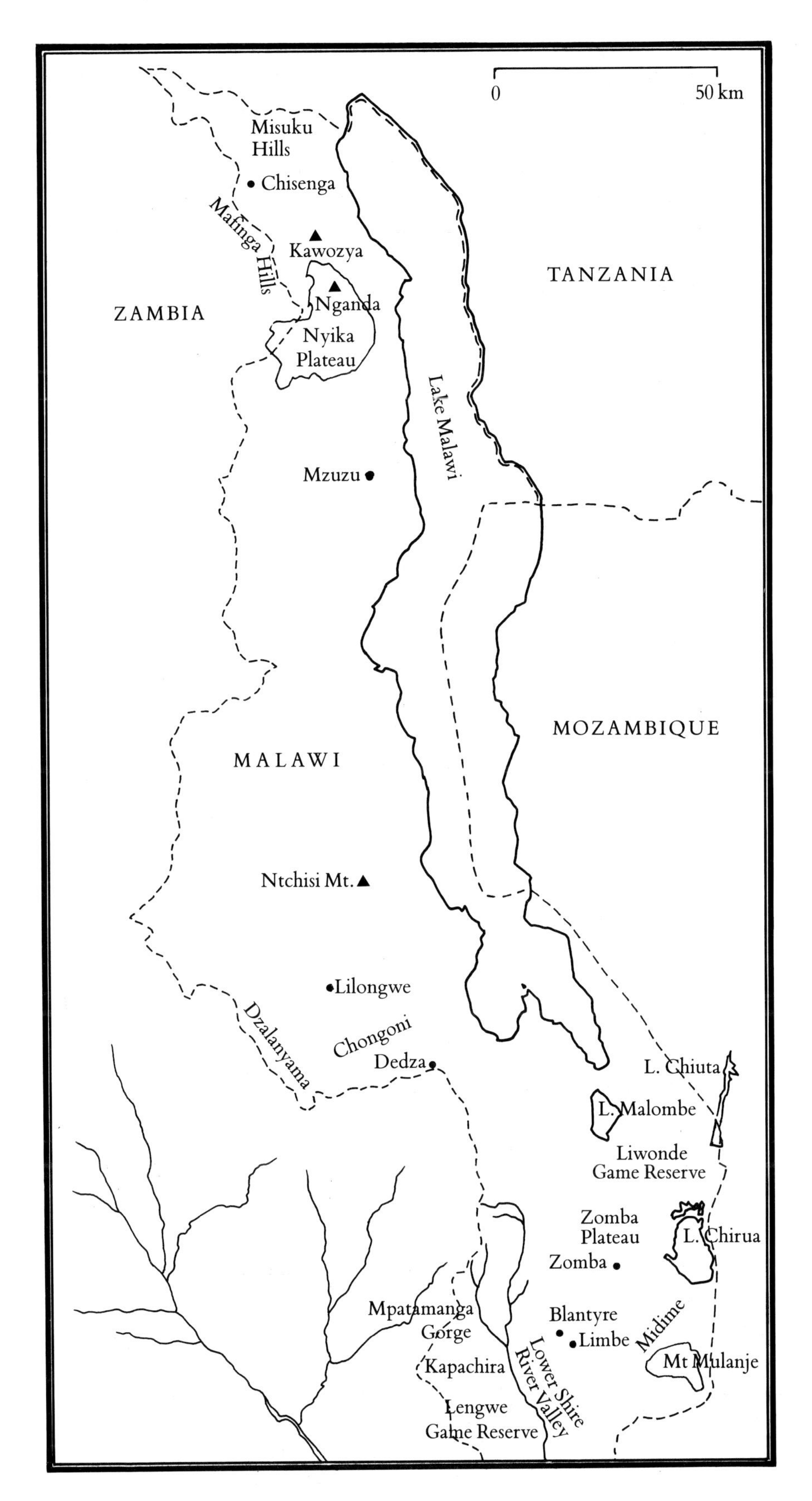

Map 4 Malawi with places mentioned in 'The Malawi connection', ch. 3.

above sea level to the rocky peaks of Mt Mulanje at about 3000 m (10,000 ft). It has been my great privilege as a botanist from Kew to visit this lovely and fascinating country 10 times in 18 years. I have been able to explore many of Malawi's well-known and little-known botanical havens, to work happily with its own botanists in the herbarium and in the field, and hopefully to play some small part in the development and establishment of its own National Herbarium and Botanic Gardens.

I first took a job at Kew in 1963, working in the Herbarium on the *Flora Zambesiaca* project which documents the plants of southeastern tropical Africa. Poring over specimens accumulated at Kew since the days of David Livingstone, I was happy enough at first to build up some picture of the flora and geography of this part of Africa from a distance, but became increasingly anxious to get out in the field and see things at first hand. Early hopes of a visit were frustrated by political problems in the region, and it was not until 1970 that the chance of a visit to Africa arose. An invitation came from Professor Blodwen Lloyd Binns, who was teaching botany at the new University of Malawi, then in the suburbs of Blantyre, to spend six months based with her. My wife of 18 months, Hilary, was invited too, and a house and transport would be provided. What more could a young botanist want?

The first expedition plans

Before setting off, we paid a visit to two old hands of Malawi botany, Jim and Betty Chapman. Jim had been a forestry officer in Malawi, or Nyasaland as it then was, in pre-independence days, and together they had had opportunities to visit forests in many parts of the country. As we got out maps and planned itineraries I felt myself infected by their love of Malawi and enthusiasm for its plants. Over subsequent years our paths have crossed again frequently, sometimes at Kew and sometimes in Malawi, and we have continued to exchange experiences and information about the flora of the country.

My first six months in Africa were probably the most stimulating period of my botanical career, and forged a strong botanical liaison between Kew and Malawi which I hope will continue to be mutually profitable for a long time to come. Apart from a very occasional lecture at the university I had no formal duties and was free to travel up and down the country collecting plants, returning to base at least fortnightly to re-stock on supplies and equipment and make tentative identifications of the collections, using the small but valuable university herbarium. I was accompanied everywhere by Hilary, who played a crucial role in organising the logistics, providing food, and acting as a field assistant and photographer. We aimed to amass a collection of dried specimens as representative as possible of the flora of the country from early February, in the middle of the rains, through to the height of the dry season in August.

Often we were accompanied by Elias Banda from the university

Hilary Brummitt relaxing at Livingstone Falls, now called Kapachira. This is the highest point in the Shire River to which David Livingstone was able to voyage uninterrupted on his Zambesi expeditions.
Photo: H J Brummitt

herbarium, from whom I acquired a great deal of field knowledge of the flora, or his colleague Hassam Patel, who enthusiastically guided us to many localities off the beaten track. We made some 4100 collections, usually in sets of four or five or more, to be deposited at Kew, at the Malawi Herbarium, and at other centres concerned with the *Flora Zambesiaca* project.

At that time Malawi had been independent for a mere four years, and one sensed the excitement of a newly emerging nation enjoying the sudden surge of development which this opportunity brought. The university had been recently established, and even though it was then housed only in a converted school at Chichiri, between Blantyre and Limbe, it had a well-qualified staff, partly Malawian and partly expatriate, and enthusiastic students. The very pleasant town centre of Blantyre seemed to symbolise the times, with its single-storey colonial offices in Victoria Avenue recently dwarfed by a new skyscraper block.

But the road system inherited from colonial days was rudimentary, and the first tar road from the main industrial and administrative centres in the south to the less developed Central and Northern Regions was still under construction. It was an expedition in itself to travel from Blantyre to Lilongwe (the present capital of Malawi), and we became used to our small car slipping and sliding through seemingly endless roadworks or being held up by rock-blasting operations near Dedza. However, if there were delays one could always wander into the woodland with a plant press, and some of our first records for Malawi were collected just at the roadside in unremarkable localities.

Mt Mulanje

The highlights of the tour, however, were the times spent on the high mountains. Although on later visits I have become very used to the stunning view of Mt Mulanje from Zomba, I still remember seeing the mountain for the first time. We drove out from Blantyre to do a little local exploring one late afternoon in February, came over a rise in the road near Midima, and were suddenly confronted by an enormous massif rising straight out of the plains in front of us, its high pinnacles catching the evening sun and seeming to tower over us as if heaven itself had been suddenly revealed. It was literally breathtaking.

Although commonly referred to as one mountain, Mulanje is in fact a range about the size of the English Lake District, a granitic intrusion rising sheer to a broad plateau at about 1800 m (6000 ft), with many rocky peaks above that to about 3000m (10,000 ft). It is uninhabited apart from a few forestry huts, and there is no road up. The ascent to this botanical heaven has to be made the hard way, on foot, as pioneered by Alexander Whyte, the first European there, in 1891.

We made our first ascent on 5 April. After making preliminary arrangements with the Forestry Department, we drove to the foot of

Nyika Plateau, in the north of Malawi, has been visited by several Kew expeditions, and a number of species new to science have been discovered there. The photograph shows the rolling grassland of the western side of the plateau, rather than the deeply dissected terrain with more extensive forests found on the eastern side.
Photo: H J Brummitt

Lobelia stuhlmannii reaches its southernmost locality on the Nyika Plateau. This is one of a number of giant columnar species typical of eastern African mountains.
Photo: H J Brummitt

the north side in early morning with Hassam Patel, who efficiently recruited a team of five porters from the local village. He gave them their standard payment of one tin of corned beef, one bag of rice, one packet of cigarettes, and 4s 6d (about 23p). After lining them up in order of seniority, he then distributed our loads between them, including large baskets of provisions which we could hardly lift and tall stacks of plant presses. It is a humbling experience to haul yourself up a steep climb of 9–1200 m (3–4000 ft) unladen, puffing and sweating, knowing that these young lads are way ahead of you, balancing your heavy loads on their heads.

The afternoon sun did its best to drain the last reserves of energy from our legs, but we gained inspiration from the changing scent as the lowland vegetation gave way to the mountain flora. Here there were numerous aromatic shrubs, such as the many *Helichrysum* species, *Buddleja salviifolia, Hypericum revolutum* and others, and patches of resin-scented Mulanje cedar (*Widdringtonia nodiflora*). We reached Tuchila hut just before sundown. As we approached, the aroma of cedar-wood smoke wafted invitingly towards us, and we were glad to find that our porters had already built a roaring fire before all but one of them set off down to their village again.

It is a different world up on Mulanji, and a botanist's paradise. We were fascinated by the flora with its eye-catching aloes (*Aloë arborescens, A. mawii*), velloziads (*Xerophyta*), whose fibrous stems are used by the local people as scrubbing brushes, the Ericaceous *Philippia, Ericinella* and *Pieris*-like *Agauria* and an occasional *Erica* itself, the grassland *Gladiolus* and red-hot pokers (*Kniphofia*), the rock-bound one-leaved *Streptocarpus* species, numerous balsams (*Impatiens*) and a host of other beautiful plants often with relatives known to us from European botany. Indeed, the predominance in places of bracken (*Pteridium aquilinum*) gave the scene a very familiar appearance for a British botanist. Most of the plants in the easily accessible parts of the mountain are well known these days, but a few of our collections later turned out to be the first ones made and some have been described as new species. One large yellow-flowered shrub had been collected once before, by Jim Chapman, and it was gratifying that our two collections of it enabled the species to be described as new (*Gnidia chapmanii*).

But Mulanje can be a dangerous place too, and we had been warned about flash floods when the rain washes rapidly off the rock slabs and fills the streams. One morning we walked out from Tuchila hut and casually crossed three small streams, stepping over them with no difficulty. At lunch time it rained, and we took shelter for an hour before heading back. We could hear the first stream before we reached

Porters carrying Dick Brummitt's equipment, assembling at the base of Tachila path to Mulanje Plateau in 1970. During 1946, in the then Nyasaland, Leonard Brass had made an outstanding collection of plants which were identified and published by J P M Brenan of Kew, including new species such as *Helichrysum brassii*.
Photo: H J Brummitt

it, and when we got there we were horrified to see a huge white foaming torrent a good 10 m (30 ft) across, carrying all before it as it thundered headlong to the plateau edge.

Already cold and wet, we thought we were stuck for a very unpleasant night out on the mountain, but our one local porter, Magombo, had experienced this situation before. He led us down the stream to where a big tree had been felled across it. To avoid encumbrance we tried to throw the day's collections in their press across to the other side, but it hit the bank and slowly slid back into the raging torrent, no doubt to descend the mountain a lot faster than we would. We had almost physically to restrain Hassam Patel from leaping in after it to rescue the specimens. Then we inched tremulously across astride the fallen tree with the foam a foot or so below our ankles, and repeated the process at the next two streams before thankfully reaching the safety and warmth of the hut. Perhaps it was fortunate that we read Laurens van der Post's *Venture to the Interior* only after that day's events, and we knew nothing of his tragedy in such a situation.

The Nyika Plateau

The Nyika Plateau in northern Malawi is a marked contrast to the rocky peaks of Mulanje. A rough road winds up through the woodland, widely known as 'miombo' after the principal tree, and, after passing the well-known stands of the very striking pink *Protea angolensis* var. *trichanthera*, opens out at around 1800 m (6000 ft) onto the broad rolling plateau grasslands. During the early rains around the end of the calendar year an abundance of flowers appears among the grasses, including many ground orchids, red-hot pokers, daisies, labiates and many other families. In the damp hollows giant lobelias (*Lobelia mildbraedii* and (*L. gibberoa*) show the affinities with the uplands of East Africa. On our first visit, in late March, the margins of the extensive road system on the plateau were carpeted for mile after mile by a lobelia of very much small stature, the bright blue-flowered annual *Lobelia trullifolia*.

Patches of evergreen forest persist here and there, usually on the banks of streams or in hollows; an occasional charred tree stump standing nearby in open grassland demonstrates that these patches are relics of a once much more widespread forest which has retreated in the face of the annual burning of the grass in the dry season. Now a national park covering some 93,000 ha (230,000 acres), the plateau is well managed, and the inevitable conifer plantations (mostly *Pinus patula*) are confined to small areas near the delightful visitors' chalets at Chelinda Camp. Animals are abundant and easily seen, including eland, roan antelope, reedbuck, jackals, hyena and leopard, but our favourites were always the zebras and it gave me some pleasure recently to name one of our collections *Plectranthus zebrarum* in their honour.

★ ★ ★

In our six months we collected in every district except Dowa. In the extreme north we were surprised at the density of population in the remote Misuku Hills, where steep slopes had been terraced for intensive cultivation, though fortunately the very interesting *Entandrophragma* forests were still relatively untouched. In Mzuzu we met up with Jean Pawek, an American missionary teacher at Marymount School, whose prolific botanical collections in northern Malawi over 15 or more years rank among the most outstanding ever made in Africa. In the Central Region we were introduced by the Pinney family to Dzalanyama in its days as a virtually untouched miombo woodland forest reserve. Population had been excluded throughout this century to create a tsetse-free buffer area on the Mozambique border, but, sadly from the botanical point of view, it has now been put to use as a cattle ranch. At Chongoni, near Dedza, we visited the Forestry Department herbarium and met A J Salubeni, who took us up Dedza Mountain. He has remained a good friend ever since, and in 1987 was the first student to enroll in Kew's inaugural International Diploma Course in Herbarium Techniques.

We also had two memorable stays at the rest-house below Ntchisi Forest, with its splendid views across Lake Malawi to the Mozambique mountains. Trips to the extreme south of the country in the Shire Valley took us into a completely different vegetation, particularly at Lengwe Game Reserve and at Kapachira (Livingstone Falls). In the heat at the low altitude being bespectacled was a major disadvantage: I could never get more than two collections into the press without the lenses of my glasses being flooded with sweat. After a few days with the temperature never below 30°C we always cooled off and relaxed with a day or two at Ku Chawe Hotel, perched on the very edge of Zomba Plateau with magnificent views of the Shire Highlands.

Nyika again

In 1972 I was very pleased to have an opportunity to return to Malawi when I was invited by Sam Kent and Hugh Synge to join a student expedition from Wye College in Kent. They had been invited to make a study of the rather inaccessible northeastern side of the Nyika Plateau, which was then outside the national park. Two students from the University of Malawi also took part, one of them David Munthali, who was assiduously collecting termite specimens at the time and is now a prominent entomologist. We camped during July and August at about 2400 m (8000 ft) to the west of Nganda hill at the limit of the road system. At that time of year it was pleasantly hot in the midday sun, but every night it froze hard and it took four pullovers and an arctic blanket to defeat the cold before we woke to hard white frosts in the morning. On the east side of the Nyika the terrain is very different from the rolling grasslands known to most visitors to Chelinda, for the country is deeply dissected by steep valleys, and forests are more extensive. We walked for a day or two at a time from base camp and pitched tents again, collecting plants and animals well away from previously collected areas.

Although it was the depth of the dry season, the altitude of the plateau meant that many species which were out of flower elsewhere could be found in flower in these steep valleys. On Kawozya peak, slightly detached from the main Nyika Plateau on the north side, we collected a composite later described as *Vernonia kawozyensis*, still known only from this one collection. Another notable plant was *Adiantum reniforme*, a fern with kidney-shaped leaves and a crazy distribution, previously known from Madeira and the Canary Islands, two localities in Kenya, then Madagascar and Mauritius, and now found for the first (and still only) time in Malawi. It was gratifying that after the expedition's report was presented to the government the Nyika National Park was extended very considerably to include the area which had been surveyed.

Later connections

Botanical developments took me back again after a tour of Zambia in 1975. In 1972 there had already been talk of creating a national herbarium by merging the university collection with the older collections of the forestry and agriculture departments, then housed separately in the Central Region, and I was involved in further negotiations about its formal establishment. The university had been moved to its present site in Zomba, where the plateau rises abruptly 900 m (3000 ft) above the town and makes a perfect backdrop for the fine new campus. Here I met for the first time Jameson Seyani, recently graduated and destined to take over the herbarium. I have enjoyed many excursions and discussions with him since our first meeting. It is largely through his dedication and vision that Malawi now has one of Africa's most active botanical institutions.

By 1977 plans were ready to be put into practice. The British Overseas Development Administration financed two visits that year to supervise the merging of the university herbarium (already including the small agriculture collection) with the historically more important forestry herbarium. It was a time for 'rolling one's sleeves up' with the herbarium staff, shifting probably some 50,000 specimens into new accommodation, writing out new folders, re-gluing loose labels, and generally creating a well-curated modern herbarium.

We had an occasional day off to get into the field to revisit places already familiar from 1970 and some new localities. Hazel Meredith was of great assistance in many ways, Cornell Dudley was glad to demonstrate the unique character of the low-altitude 'mopane' woodland in the Liwonde Game Reserve, and his fellow American, Bruce Hargreaves, would drop everything to take one to see his favourite succulent euphorbias. And on the forestry side, Pat Hardcastle and Rodney Nkaonja were always most cooperative and encouraging.

On follow-up visits in 1979 and 1980 the curatorial reorganisation of the herbarium was completed and it was time to try to achieve a

Chisenga rest house, 1970, with the Mafinga Hills beyond. Sadly, this charming little rest house is no longer in use and has fallen into disrepair. These hills, on the Zambian border, although not particularly high, support a surprising number of endemic species.
Photo: H J Brummitt

proper administrative structure for its staff. Although the collections were then functioning as a valuable national reference resource for many people working with plants in Malawi, and were increasingly visited by taxonomists from other countries, the staff were drawn partly from the university and partly from the Forestry Department. Organisation and financing presented many difficulties.

Steve Blackmore, now at the British Museum (Natural History), was then on the university staff and in charge of the herbarium. We had several top-level meetings with government and university people in the hope of setting up a single new body, independently financed but affiliated to both 'parents'. However, although goodwill was forthcoming on all sides, the lack of a precedent for such a body seemed to be a stumbling block, and the project was habitually referred back to yet another committee.

In 1982 a Kew expedition gave the opportunity for another return to Malawi. Roger Polhill, Nigel Hepper and I were greeted in Zomba by Jim and Betty Chapman. A few glorious days on Mt Mulanje preceded a tour to the north region with Roger Polhill and my old friend Elias Banda. How I enjoyed the new tar road, which then extended well north of Lilongwe! The only hindrance to progress was Roger's ability to spot a mistletoe in a treetop while travelling at 70 mph, and since misteletoes were one of the major objectives of the expedition we had to slam the brakes on every time.

Perhaps climbing the Mafinga Hills from our camp site at the foot near Chisenga was the highlight for me. And on our disastrously late arrival (those mistletoes were too common in the north!) on the Nyika Plateau hours after sundown we were delighted to be greeted by Françoise Dowsett-Lemaire, Belgian ornithologist-turned-botanist, who with her husband, Bob Dowsett, was conducting an intensive multidisciplinary study of the evergreen forests. On our return to

Zomba, further discussions on establishing the national herbarium were arranged, stimulated by our arrival from Kew, but it was frustratingly obvious that nothing had changed since my discussions two years before.

Back again in 1984, after a tour of Zambia collecting *Protea* species with my Zambian friend Sylvester Chisumpa, I found Jameson Seyani firmly in charge of the herbarium after his return from Oxford with his D.Phil. My arrival again prompted a flurry of meetings, and a new element emerged. The town of Zomba has boasted a botanic garden since the 1890s, when Alexander Whyte, working in Sir Harry Johnston's administration, planted many exotic species on a site below the residency (now the government hostel). These early plantings were still there and well cared for, but the area had become little more than a public park. Jameson Seyani had great visions for extending and revitalising the botanic garden as a modern scientific institution, including the national herbarium. We took his ideas to the Vice-Chancellor of the University, Dr David Kimble, whose forthright support was greater than we had dared to hope for. After a brief pause for reflection he concluded, in matter-of-fact voice, that Malawi did not want one botanic garden but three. If Zomba had one in the South, it was necessary to develop one in Lilongwe for the Central Region and another in Mzuzu for the North!

In addition to the enlightened and far-sighted attitude of the Malawian government, much of the credit for getting an appropriate constitution for the Herbarium and Botanic Garden drawn up and the necessary bill passed through Parliament must go to Jameson Seyani. After the many setbacks and delays it took all his drive and determination to bring the project to full fruition. At a late stage I was able to send him a copy of the National Heritage Bill, by which Kew was set up under trustees in 1985, and this acted as a catalyst for the legal people in Malawi to prepare similar legislation. Returning for a brief visit in 1986, I hoped to hear that Parliament had passed everything, but, alas, a last-minute hitch meant that some redrafting was necessary. At last in 1987, some 15 years after the first discussions I received a telephone call at Kew from Jameson in Zomba to tell me that the bill had finally gone through and the National Herbarium and Botanic Gardens of Malawi were formally constituted, a significant milestone in the co-operation between our respective institutions.

4 Searching for a forest coconut in Madagascar

JOHN DRANSFIELD

From 1980 to 1986 Dr Dransfield was involved in a joint project with Dr Natalie Uhl of Cornell University in the USA to produce a *Genera Palmarum* – that is, an account of the world's palms at the generic level. The palms are said to be second only to the grasses in economic importance and of very great scientific interest. As they worked towards the completion of that volume, they came to realise how little was known of the palms of the island of Madagascar, the Malagasy Republic. Sufficient was known however, to indicate that they were both numerous and peculiar; yet it was not at first at all easy to do fieldwork on the island. This chapter tells the story of an expedition there, which helped to encourage an increasing collaboration between Kew and Madagascar.

In 1984 an American palm enthusiast, Mardy Darian, sent to Natalie Uhl and I the hard woody shell or endocarp from the fruit of a palm from Madagascar. The endocarp had been given to Mardy by a Malagasy palm enthusiast, Gerard Jean. It was said to be from the fruit of a palm native to Madagascar and known locally as *voanioala*, literally the 'forest coconut'. The palm was said to resemble a coconut in habit and to inhabit extremely remote areas of forest in the northeast of the island. No more information was available.

The endocarp did not match that of any known native Madagascar palm and with its three 'eyes' it clearly came from a palm belonging to the Tribe *Cocoeae*, a Tribe (group of genera) otherwise unrecorded for the island. In fact the hard shell strongly resembled that of a member of the group of palms we include in the Subtribe *Attaleinae*, that is, the babassu and cohune palms and their relatives, wild oil-palms confined to South America with a few outliners in Central America and the Antilles. The endocarp was about 9 × 5 cm (3.5 × 2 in), very thick, deeply grooved and with three deeply sunken pores (the 'eyes') at one end. As it was already old, it seemed pointless to try to cut it open.

At first I was inclined to suggest that the palm might be a new record for an otherwise entirely New World Subtribe in Madagascar. Then I began to think that that was perhaps too improbable and that the endocarp might have been collected from a true babassu-type palm introduced to Madagascar during colonial times and since forgotten. However, Mardy Darian insisted that the palm was a Madagascar native. As *Genera Palmarum* went to press, we included in the manuscript a statement to the effect that there might well be a member of *Attaleinae* native to Madagascar. How wrong we were!

Arrival in Madagascar

In early October 1986 David Cooke and I from Kew, Pete Lowry and Jim Beach from Missouri Botanical Garden, Armand Rakotozafy from Parc de Tsimbazaza in Tananarive, Marion Nicol from Britain, Hilary Simon and Paul Dark from Yale University and Nick Lindsay from Jersey Zoo, all descended on Maroantsetra at the head of the Bay of Antongil in northeastern Madagascar to do fieldwork on the Masoala Peninsula. The first six of us were there to do botanical collecting, and the last three to do zoological and anthropological work. Rarely before

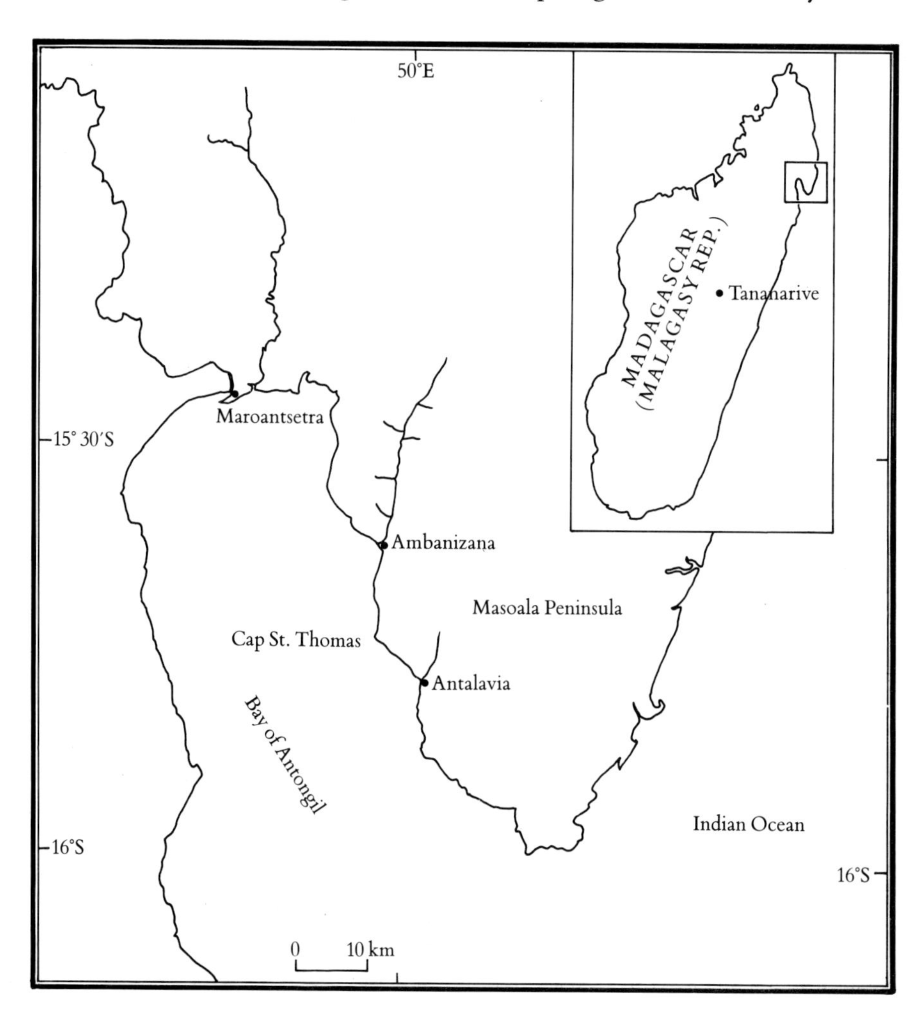

Map 5 Portion of Madagascar with localities mentioned in ch. 4.

The Bay of Antongil on the east coast of Madagascar is bounded by the Masoala Peninsula, still covered with rainforest and very little explored botanically.
Photo: J Dransfield

have I worked with such a large crowd and I was slightly apprehensive that we might get in each other's way. However, in the end I can truly say that I have never enjoyed fieldwork so much as with this multidisciplinary crowd.

Shortly after our arrival we contacted Gerard Jean on the advice of another palm enthusiast, Dominic Halleux. Gerard was to join us for the next three weeks and he shared his extensive knowledge of Madagascar palms with us. As soon as we met we started to talk about the 'forest coconut'. Where did it grow? Was it really a native? Was there any chance we might get to see it? A native it certainly was, replied Gerard, but it grew a long way from Maroantsetra and we might have to walk for at least three days into the rain forests of the interior to reach the nearest colony. This would mean a trip of at least seven days, and we would not be able to do any palm collecting on the march as we would have no space for the specimens or, indeed, time to collect.

It was gloomy news and I began to think that maybe we would have to pay a villager or Gerard himself to go off into the interior to obtain specimens of the palm. Privately I also felt that perhaps Gerard was exaggerating the distances and the difficulties, and that after he had seen how far we could walk, the number of days required to reach the 'cocoid' might be decreased.

We went to the Masoala Peninsula by boat from Maroansetra to the tiny settlement of Hiaraka. Here we busied ourselves in the forest, and I became acquainted for the first time with many wonderful palms which I had known only from miserable scraps of herbarium material. The terrain was, for me, quite tough and I realised that I would have been well employed before I left Britain in trying to get fit. We had the great pleasure of finding a new species of *Louvelia* and another of *Halmoorea*. We also saw beautiful undergrowth species, such as *Neophloga thiryana* and *Dypsis humbertii*, as well as majestic species of *Ravenea*, *Orania longisquama* and several as yet unnamed tall species of *Neodypsis*. Of the 'forest coconut' there was no sign or rumour. However, Gerard

A typical village house on the shores of the Bay of Antongil. The walls are constructed from the leaf-stalks of the *Raphia* palm while the thatch is made from the *Vonitra* palm leaves.
Photo: J Dransfield

Gerard Jean (on the left) with François and Joubert, the two villagers of Ambanizana, who led the Kew team to the 'Forest Coconut'.
Photo: J Dransfield

assured us that the next important settlement along the coast, Ambanizana, would make a good base for striking off into the interior to look for the palm.

So, after a week at Hiaraka, we moved by boat down the misty, forested coast of Masoala to Ambanizana, where we arrived in the evening after a rain shower. We reported to the village headman, and as long flowing speeches were intoned in Malagasy I gazed around the village to the surrounding forested mountains and wondered what palms were lurking there, waiting to be discovered. Gerard worked hard into the night, talking to villagers and asking about the possibilities of reaching a colony of the 'forest coconut' from our camp.

The next morning after breakfast a plan was hatched. Apparently two villagers, François and Joubert, had seen the palm growing inland from another settlement still further down the coast and it should be possible for us to reach that colony. However, the *Beatrice*, our boat, had already returned to Maroantsetra, so we would have to reduce the size of the party to the number of people who could be accommodated in a dugout canoe. It was decided that the party would consist of François and Joubert, Gerard, of course, and David and me. The plan was to leave by dugout at three o'clock the next morning while the sea was still calm and to paddle for three hours down the coast to Antalavia. We would then walk inland to the 'forest coconut', make specimens, have lunch and walk back, arriving just before dark, with time to paddle the three hours back to Ambanizana.

We were advised to take the minimum of personal luggage and the bare essentials for collecting (i.e. a folding saw, David's climbing irons, secateurs and labels, and one camera between us). Gerard warned us that if we stopped on the way to look at other palms we would ruin the schedule – he had already seen how easily I can become distracted by other specimens.

Somewhat chastened by this plan, David and I decided we would spend the rest of the morning on a gentle stroll through the nearby forest and the afternoon getting our equipment ready for the three o'clock start. At dark we retired to our tents in a state of excitement. I tossed and turned that night, waking up at frequent intervals to check the time. At three o'clock I was up and dressed, only to find that the dugout had not arrived. At dispiriting hourly intervals I emerged from my tent to find that it still had not arrived. Finally, just after dawn, the tiny dugout appeared round the headland into our bay. How could we possibly stick to our plan if we were so late in starting?

As the sun rose above the mountains we squeezed into the canoe, and amid much shouting of messages, largely valedictory, the five of us slid away from the beach on a flat calm sea. The canoe seemed crammed to bursting with bodies and luggage, and after 20 minutes every muscle in my body was either numb or screaming from sitting in awkward positions. Each time I moved to ease a buttock the canoe swayed ominously. After a while it became obvious that it was going to take us a good deal longer than three hours to reach Antalavia, if the position of this tiny settlement was correctly marked on the map.

Eventually I became somewhat inured to the pain and numbness and found the peace of mind to enjoy our situation. Here we were, skimming in a tiny dugout on the calm Indian Ocean over the most exquisite coral, in sight of the forested shoreline of one of the most wonderful islands. Great trees of *Barringtonia, Terminalia, Intsia* and *Sideroxylon* hung over the beach, festooned with epiphytes. Occasionally we would see and smell curtains of the climber *Stephanotis grandiflora* in full white flower. From time to time we would pass villagers out fishing from their dugouts and shout greetings over the water.

As we rounded Cap St Thomas, the waves increased somewhat and the journey became slighly more tense. At about 10.30 a.m. we pulled into the shore to stretch our legs and to cut a branch of a tree, which it was my duty to hold upright to act as a sail. I am convinced this hindered rather than helped us and only added to the number of muscles being strained. Eventually at about 11.30 a.m. we caught sight of the river mouth at Antalavia, and about six hours after leaving Ambanizana we were gently easing our way over the sand bar into the quiet lagoon at the mouth of the river.

David and I were in a state of exhaustion and I wondered what Gerard was planning to do next. We pulled up opposite a small fish smokehouse and clambered ashore, sat in the shade and let events develop. It was quite out of the question to go chasing palms in our state. A new plan was concocted. At five o'clock next morning we would leave the hut and walk into the forest, collect the cocoid and return. Then, after dinner, we would leave in the dark to paddle back to Ambanizana.

What a blessed relief not to have to move a screaming muscle again that day! David and I whiled away the rest of the afternoon in the shade of a grove of bananas, dozing or idly watching the activities of the family living at Antalavia. A spring behind the house provided clean water for cooking and bathing. Visits to the non-existant toilet involved a canoe-ride across the lagoon, and before I retired to my sleeping bag that night I was paddled there, feeling very much like a dog being taken out for an evening walk!

Related to the commonly cultivated stephanotis (*S. floribunda*), *Stephanotis grandiflora* forms curtains of fragrant white flowers in the forest along the shores of the Bay of Antongil.
Photo: J Dransfield

Other palms, but where was the forest coconut?

That night the weather was kind: there was a clear and star-spangled sky and a gentle sea breeze kept the temperature pleasant. At four o'clock we were up and getting dressed. Gerard, François and Joubert cooked rice while David and I stuffed ourselves with as much muesli as we could, filled our water-bottles with sterilised water and pulled on our jungle boots. Just as it was getting light we piled into the canoe to paddle to the head of the lagoon and begin the walk into the forest.

The river at Antalavia crashes out of the hills through a deep gorge, noisily tumbling into the lagoon. We left the canoe at the first waterfall, climbed up a steep slope and followed a contour track up the gorge for

about 1 km (⅔ mile). Eventually we found ourselves back down in the bottom of the gorge and forced to cross, immediately getting our feet wet. They remained wet for the rest of the day. The path skirted the riverbank, then again we had a steep climb out of the gorge onto a high ridge top, which we decided to follow to avoid the fearsome waterfalls we could hear down below us. Here at last we were in undisturbed primary forest full of palms.

Mindful of Gerard's instructions not to look at any palms on the way, I could not help noticing how rich they were. One of the commonest big palms was *Neodypsis lastelliana*, reaching the forest canopy and conspicuous with its warm reddish-brown crownshaft and elegant pendulous leaflets. Here is a palm which deserves to be widely planted. It has all the elegance of *Oncosperma tigillarium* but without those fearsome spines. We also could not help noticing our un-named *Louvelia*, which appears to be common all the way down the coast of Masoala. I also saw a large rather untidy *Ravenea*. *Vonitra fibrosa* was common everywhere, particularly along the river. However, a small elegantly fibrous *Vonitra* halfway up the ridge was certainly not *V. fibrosa* nor any of the other species in the *Flora*. Probably a new species, it was the first of the many palms I encountered on this trek which there was no time to collect.

We followed the path and slithered back down into the valley bottom, listening to the thundering waterfalls we had avoided. The next stretch of the valley was both beautiful and treacherous. The valley bottom was choked with huge boulders covered with moss, ferns, orchids and *Impatiens*. Our path occasionally skirted the bottom but for most of the time we had to hop gingerly from boulder to boulder. We were very conscious of the deep holes, partially covered

(*opposite page*) The path to the 'Forest Coconut' led up a river choked with slippery boulders.
Photo: J Dransfield

David Cooke climbs one of the two trees of the 'Forest Coconut' (*Voanioala gerardii*) to make the first herbarium collections of this new genus.
Photo: J Dransfield

The young fruit of *Voanioala gerardii* are similar to those of other members of the group to which the coconut belongs.
Photo: J Dransfield

with ferns, and from time to time we could hear the river tinkling deep below us.

After about an hour of concentrated effort we reached a small clearing with the remains of a jungle camp. Here grew a squat massive palm which I believe to be the poorly known *Antongilia perrieri*. We were allowed five minutes rest. I snapped a few pictures and earmarked the *Antongilia* for collecting on the way back. As it happened, we had no means of carrying additional specimens on the way back and sadly this palm will have to wait for another visit. At least there is already good material of this species in the Bailey Hortorium, collected by the palm specialist, the late Hal Moore from the east coast of the Masoala Peninsula.

Above this clearing the path wended its way through a further boulder field. We passed under *Halmoorea trispatha* and an unidentified species of *Neodypsis* while in the undergrowth we saw *Neophloga thiryana, Dypsis humbertii* and *D. mocquerysiana*. Eventually we emerged from the boulder-choked area into a gentle upland valley with the river merely a trickle meandering through fine forest rich in tall pandans and an almost unbelievable abundance of palms.

By now it was after ten o'clock and David and I were both rather tired. I was unwilling to give voice to my flagging spirits as we trudged on, alternatively wading through the river up to our thighs and slithering up out of the riverbed onto the banks. But as David fell in yet again, he let fly, 'I hope you d— well think this is worthwhile'. I hardly dared reply and continued trying to keep up with the guides.

Perhaps they too were beginning to feel we should have reached our destination, for a short time later I was forced to follow them up a steep slope to look at a dead palm which someone had cut down to obtain the cabbage. This, they said was the 'forest coconut'. I knew we were looking at a species of *Ravenea* but I was in no state to argue. Now I was quietly convinced that we were on a wild-goose chase.

But suddenly things began to look up. We came into an idyllic area of swampy valley bottom, filled with palms and pandans. Among the palms were two I had not seen before. David and I were now trailing rather badly and, although I was frightened we would be left too far behind, I had to look at the palms more closely. One was a huge tree-palm with immense curled leaves, the sheaths covered with pinkish grey hairs; it looked a bit like a species of *Orania* but none I knew. There were no signs of flowers or fruit or even of dead inflorescences, and to this day I still don't know what that beauty was. Again, I shall have to wait for the next trip. The other palm was a low squat palm with a crown of many leaves rather like a shuttlecock. What on earth was this?

Before I had time to look more closely for flowers or fruit, François and Joubert re-emerged ahead of us, clutching a handful of big palm fruits. Immediately I recognised the fruit Mardy Darian had sent, and I knew that we had 'made it'. Forgetting the squat palm, I let out a whoop of joy and with renewed enthusiasm we pounded through the forest to see the tree. Five minutes later we stood under a large

Symbol of Madagascar, the Traveller's Palm or Traveller's Tree, *Ravenala madagascariensis*, is not a palm but a relative of the bananas and bird-of-paradise flower (*Strelitzia*). It is found throughout the island and is intensively used for thatch, walling and occasionally for food.
Photo: J Dransfield

tree-palm with a conspicuously stepped trunk and massive crown of leaves, the ground beneath carpeted with empty endocarps and seedlings. Gerard quickly organised us. He, Joubert, David and I would collect specimens while François would return to the river to cook rice for lunch.

David strapped on his climbing irons and climbed up into the crown of the palm. Here he roped himself onto the younger leaves, and, to shouted instructions, began to cut inflorescences in bud and young infructescences. Then he removed leaves at the point of their attachment to the trunk. In the meantime, dodging falling leaves, I wrote extensive field notes and chopped up the leaves, taking samples of the basal section, mid portions and leaf tips and numbering each fragment with a hanging jewellers' tag, I then attacked the inflorescences and infructescences, reducing them to a manageable size. As David finished and climbed down the trunk safely, I scrambled around, slithering on the steep slope to pick up handfuls of fallen fruit, some to take back to Kew to grow, others to add to the herbarium material.

About an hour after starting the task of collecting this splendid palm, we had finished – a rushed job, which I would have preferred to do in a more leisurely way. We filled woven fertiliser sacks with the specimens and then returned to the river. François had prepared mountains of steaming white rice, which we gobbled down with tinned sardines – a memorable and tasty meal.

Triumphant return

Then began the journey back. We were now laden with booty and could not make fast progress, but slippery boulder after slippery boulder we made our way back down the river valley. I took hurried looks at the mystery palms we had passed on the way up, wondering if I would ever return to see what they were. As we made our way down the last steep slope to the lagoon, a party of red-ruffed lemurs screamed and chattered at us in the tree tops. We reached the smokehouse shortly after five o'clock, with just time to bathe before dark.

Rather than attempt to paddle back that night to Ambanizana, we decided to leave early the next morning. So, having filled ourselves with rice and fish, we retired to our sleeping bags. I lay awake for a long time, thinking about the wonderful palm we had just collected. It was clearly a new genus and also much more closely related to the coconut than to the babassu palms. That made it the first member of this group of palms to have been found in Madagascar and obviously a very rare palm indeed. What of its natural history, what could pollinate the flowers and what animal could possibly disperse the large and heavy fruit? I felt then, and still feel now, that the fruit is too large to be dispersed by any extant Madgascar animal. As I dropped off to sleep that night I wondered whether the fruit might not have been adapted to dispersal by the giant extinct elephant bird, *Aepyornis*!

At three o'clock next morning Gerard woke us and after a quick drink we packed up our belongings, loaded the canoe and were soon easing our way into the Bay of Antongil. A dying moon cast a faint light, highlighting a large thundercloud over the mainland of Madagascar. Progress was slow, but now it didn't seem to matter. We had been completely successful. As it became light we saw large marine turtles surfacing and diving near our canoe. At Cap St Thomas the water was more choppy than on the way out, and to rest we pulled in and had breakfast. Then it was another two and a half hours paddling back to Ambanizana. At our camp we spent the rest of the day easing muscles and preparing the specimens, pressing leaves and wrapping up the flowers and fruit, eventually packing everything in polythene and dousing it in alcohol – and, of course, giving everyone else the details of our exploits.

Two years later back in Kew, the palm has been named *Voanioala gerardii* and is causing considerable excitement. From the seeds brought back have grown seedlings, and from the roots of the seedlings, counts have been obtained which seem to show that this palm has the highest number of chromosomes of any known flowering plant.

5 Tortoise islands in the Indian Ocean

S A RENVOIZE

During the last three decades several Kew botanists have visited and studied various islands in the Indian Ocean, including Sri Lanka, Mauritius, Réunion, Seychelles and Aldabra – the latter being investigated by a multidisciplinary expedition.

Aldabra is a remarkable coral atoll situated 420 km (260 miles) northwest of Madagascar and 640 km (398 miles) from the East African mainland. Movements in the Earth's crust many thousands of years ago elevated the surface of a coral reef above sea level by more than 100 m (328 ft) resulting in a large island with sufficient altitude to provide suitable conditions for the development of a variety of habitats. The elevation was followed by a rise in sea level and subsequent breaching of the outer rim to form a lagoon. The atoll is now 8 m (26 ft) above sea level but still comprises a considerable area – 15,500 ha (38,300 acres). It is composed of four main islands, which form an oblong ring 34 km (21 miles) long and 14.5 km (9 miles) across at the widest point. Of all the coral atolls in the world, Aldabra is the largest in terms of terrestrial area.

The giant tortoises which abound on Aldabra, especially at the eastern end, feed early and late in the day and normally shelter from the sun under trees and bushes at other times. Aldabra is the last Old World natural refuge for the giant tortoise which has been exterminated on other islands in the Indian Ocean.
Photo: S A Renvoize

The variety of habitats which have evolved on Aldabra support an interesting, large and diverse flora and fauna, the species of which have either reached the atoll by long-distance oceanic dispersal from neighbouring islands and continents or, in many cases, have evolved from early colonisers into unique forms. The general topography is saucer-like, with an outer rim providing protection from adverse weather conditions. The more commonly occurring low-lying, unprotected, sand-covered coral reefs found elsewhere in the world have limited habitat diversity, and their plant and animal populations are periodically depleted by hurricanes and tidal waves.

Apart from its exceptionally large size and varied flora and fauna, Aldabra is also the last natural refuge in the Old World for the giant tortoise, a species which was once widespread over the high islands of

Map 6 Aldabra atoll with place-names mentioned in ch. 5.

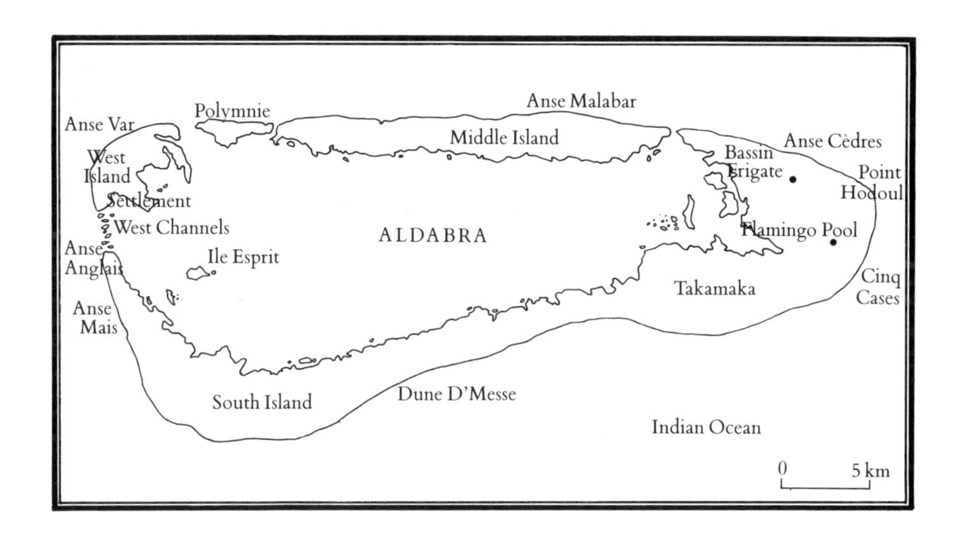

the western Indian Ocean. At the present time an enormous population of these herbivorous reptiles is supported on the atoll.

The isolation of Aldabra, its complex geological history, diverse habitats and varied wildlife make it a locality of particular interest to biologists. As a subject for scientific research its value is immensely increased by the fact that the environment has not suffered in any serious way from the impact of man. This makes Aldabra an unusual place in the world today.

Oceanic islands are very sensitive to disturbance: their environments often hold a delicate balance of nature and when this balance is upset the results may be disastrous. In the past it was the frequent habit of seafarers to introduce rabbits, goats and other domestic animals to remote islands, thus ensuring a potential supply of fresh meat at the various landfalls made during their long voyages. The legacy of this habit is numerous islands which once possessed unique floras now stripped of their vegetation by the depredation of alien animals. This has often led to serious erosion and in recent years has been further exacerbated by timber felling and mining activities.

The fact that Aldabra has survived the fate of so many islands is undoubtedly due to its remoteness, away from traditional trade routes between India, Madagascar and the east coast of Africa. Added to this is the inhospitable terrain, which is mostly covered with impenetrable scrub-forest, and the absence of a permanent supply of fresh water.

Aldabra was probably known to Europeans by the 16th century and may have been visited before that time by Arab and Chinese seafarers. It has never had a permanent settlement, although small groups of people have resided there intermittently for the past 100 years.

Following the extinction in the Seychelles and Mascarene islands of rare land birds by the end of the 18th century and of giant tortoises by the mid 19th century, the unique scientific value of Aldabra began to be appreciated. When in 1888 the atoll was leased for the development of coconut plantations and timber felling there was an outcry from the scientific community, who protested to the government and appealed for protection of Aldabra and its tortoise population. Amongst the protesters was Charles Darwin. In spite of this opposition the development went ahead; fortunately it, and subsequent commercial activities, remained on a small scale, causing no serious damage to the atoll's ecosystem.

Before the 1960s scientific studies on Aldabra had been fragmentary, starting with visits at the end of the last century and continuing intermittently up to 1967. The visits were mostly by naturalists studying specific animals or plants.

Threat to the natural history

In 1966 the British government revealed plans to develop Aldabra as a military base. With such a threat hanging over the atoll, it was clear that a major research effort was needed. The Royal Society of London and the Smithsonian Institution of Washington were two of several organisations to protest to the government and, as an immediate response to the threatened destruction of the atoll's wildlife, they organised an expedition to gather as much information as possible before the bulldozers moved in.

After a short reconnaissance visit in 1966, a large multidisciplinary expedition was launched in 1967. Although originally planned to start in August 1967 and end in April 1968, it was extended by almost a year to February 1969. It consisted of six phases, each phase coinciding more or less with the wet or dry seasons. At the time of planning the expedition Kew was asked to provide a botanist to study the vascular plants of the atoll for the preparation of a *Flora*; I was selected for this task. Over the 18 months of the expedition 40 scientists spent varying amounts of time on Aldabra.

First impressions of the atoll

I arrived on Aldabra with four companions after a rough sea voyage from Mombasa in a small trawler. When we sighted the atoll on the morning of the third day the sea was calm and before our eyes was a thin pale-green line on the horizon, which gradually materialised into a coconut plantation and *Casuarina* forest. Our landfall was opposite the small settlement at the western end of the atoll. After tossing about on the boat it was bliss to put our feet on something which wasn't moving. We went ashore at low tide, which necessitated a walk across the wide exposed reef-flat.

Although this was no great distance, it nevertheless took us a long time as we could not resist peering into all the tiny pools, each populated with marine creatures which I had previously known only from textbooks. It was also my first experience of tropical plants, albeit very strange ones: the sea 'grasses'. These plants, with tufts of straplike leaves growing up from slender rhizomes, form underwater swards where sand covers the flat reef. The sea grasses are some of the few

Steve Renvoize getting down to look for small grasses in a patch of sandy turf at the eastern end of the atoll.

flowering plants which grow completely submerged in sea water. Usually they are found in lagoons or fringing reef-flats as, like most flowering plants, they depend on sunlight for photosynthesis and can only survive in shallow water.

The expedition was based in the settlement on West Island. Five members were already there when joined by ten of us. We were accommodated in various wooden huts, made available by the small community of Seychellois labourers and their families who worked under contract to dry fish and collect coconut fibre. From this base the members of the expedition dispersed to temporary camps set up around the atoll where they could pursue their research.

The various camp sites around the atoll were located according to a supply of fresh water from wells. These wells dated from past occupation and some may have a long history associated with turtling activities. They were usually a hole in the flat rock, sometimes covered with a slab or protected from animal pollution by a low surrounding wall. The source was a fresh rainwater aquifer lying on top of the saline water which permeates the lower levels of the rock. Occasionally the well water proved unpalatable but most of the time it was perfectly acceptable, in spite of the occasional occurrence of small fish and tiny crustaceans.

During my three-month stay on the atoll, from January to March 1968, I was in the company of ornithologists, herpetologists, entomologists and marine and freshwater biologists. Each day we would disperse to our various study sites or follow planned routes, recording and collecting. In the evening we would gather over our reconstituted dehydrated potato, freeze-dried peas and tinned steak to share our experiences and finds for the day.

My plan was to collect plant specimens from as many localities as

Several camps were set up by the expedition and a permanent research base was established in huts. This camp at Cinq Cases shows the rough coral limestone rock which is exposed over much of the atoll.
Photo: S A Renvoize

possible in order to assemble a comprehensive collection of all the different species growing on Aldabra. I also wanted to assess the distibution and frequency of the species. With the information gained from my collections, a *Flora* was written which catalogued and described all the plants of the atoll; this work has been useful to other researchers studying insects, birds, tortoises and the general ecology of Aldabra.

As a result of the collections which I made, along with other members of the expedition and past visitors to the atoll, a total of 274 vascular plants were recorded. Of this number, 87 species are clearly introduced, including coconut, tamarind, sisal, date palm, cotton, lime, bottle gourd, capsicums, tobacco, tomato, lemon grass, various pot herbs and numerous well-known tropical weed species. These introduced plants were usually restricted to a few individuals or a small localised population and in no way threatened the natural plants of the atoll. Of the 187 natural plants of Aldabra, 42 of them, 20 per cent of the total, are endemic. This is a remarkably high proportion for a relatively low island but is probably a reflection of the very long period – 80,000 years – that the island has been available for plant colonisation, and the consequent evolution of new endemic species from the isolated early colonisers.

From the moment you set foot on Aldabra you are aware of a unique environment, where the plants and animals live a harmonious existence which is seldom disturbed. The tortoises have no natural enemies and many of the land birds are without fear of man. All the land animals ultimately depend on the plants for their survival, eating either the leaves or fruits and seeds. In some cases they may restrict their diet to just a few species, as is the case with some of the birds, or they may eat almost any plant which is accessible, as do the tortoises.

The settlement area occupied a narrow strip behind the beach and was entirely covered by a coconut plantation, with the inevitable introduced weeds growing around the huts and various vegetables brought by the seychellois to add variety to their diet of rice and fish. The major part of West Island, however, is covered with natural vegetation, which gave me an immediate opportunity to become acquainted with the indigenous plants of the atoll, but I did not stay at the settlement for long. Soon after my arrival I made an excursion to the extreme eastern end of the atoll.

Moving around Aldabra is not easy; the principal mode of travel was by inflatable boats across the lagoon. However, as the lagoon drains almost entirely at low tide through the various channels between the islands, boat journeys had to be carefully planned to be sure of sufficient water. More than once boats were left high and dry and the occupants obliged to sit and wait several hours for the tide to come back in to complete their journey. Travelling around the outside of the island was hazardous due to the fringing reef and generally rocky coastline, which is continually pounded by rough seas. Only in a few places are there small beaches where a landing can be made, although it is often difficult.

Rubber boats were often used to travel between islands of the Aldabra Atoll during the multi-disciplinary expeditions sponsored by the Royal Society, London, and the Smithsonian Institution, Washington, following the threat during the 1960s to build a military airfield there.
Photo: S A Renvoize

Variety of vegetation types

The eastern end of the atoll supports the greatest variety of vegetation types. Here the South Island, which is the largest island of the atoll, reaches its greatest width and has the greatest diversity of land forms. On the coast there are sand dunes and rugged coral limestone; this forms a high rim which slops downwards inland to a large area of flat limestone slabs and shallow basins. The sand dunes are largely unstable with a few trees or shrubs on their leeward sides; where the sand had blown more widely, it is often stabilised by an extensive cover of grasses. Where the rugged rock is exposed, small shrubs and herbs established communities in the sheltered crevices and numerous 'pot-holes'. The inland flat area was of particular interest as it is here that mixed scrub forest has developed. This mixed scrub is replaced further inland by impenetrable *Pemphis* thicket; the vegetation finally changing to mangrove forest on the fringe of the lagoon.

The mixed scrub forest is the most interesting plant formation on Aldabra; its physiognomy includes savanna-type formations, with scattered trees and a ground layer of grasses, sedges and herbs, and dense extensive thickets or small groves where the trees grow noticeably higher than elsewhere and where some of the rarest species on the island are found. It is in the mixed scrub forest that most of the plants endemic to Aldabra are found. It comprises the greatest variety of species and occurs in several different forms according to topography, soil type, availability of fresh water and degree of shelter from adverse weather conditions, especially strong winds and salt spray.

The mixed scrub includes many species bearing small soft fruits, which are an ideal source of food for several different types of land birds found on the atoll. The variety of fruits and seeds is diverse enough to fulfil the selective requirements of the different species. The three native figs provide much of the diet for the Comoro Blue Pigeon, which is frequently seen in the tops of the trees; in contrast fallen fruits and seeds provide food for the ground-loving Malagasy turtledove. The Malgasy bulbul feeds on a variety of fruits from *Scutia, Flacourtia, Apodytes, Scaevola, Solanum* and *Passiflora* and the tiny sunbird supplements its diet of insects with nectar from the flowers of *Euphorbia abbottii* and *Lomatophyllum.*

Apart from providing a direct source of food for many birds, the mixed shrub is also important in their nesting habits. The trees and bushes provide cover and a varied source of nesting material. The delicate grasses, fine roots and fibres, dry papery leaves and plant tendrils all provide soft lining materials whereas coarser twigs, leaves and roots are used in the main structures. The mixed scrub provides food and shelter not only for many of the birds on Aldabra but also for the tortoises. The major part of the tortoise population is located at the eastern end of the atoll. Here they can browse the lower leaves of the shrubs and trees or graze the herbs which grow under the trees; 20 species of trees and shrubs are eaten by tortoises.

Movement by foot around the southeastern corner of the atoll was not too difficult, either along the coastal sand dunes or by threading a way through the more open areas of mixed scrub. It was, however, easy to get lost inland; there were no hills for orientation and the trees were mostly slender and less than 5 m (16 ft) high, added to which they were often covered with ants and impossible to climb. Regular routes were marked with cairns, since they were often across bare rock; even where there was plant cover, the use of the route was too light for a path to develop through wear and tear of the turf. When travelling away from the standard routes, it was necessary to maintain direction by the use of a compass or by marking the trees. If this failed then the only salvation was to walk to the coast, following the sound of the waves, and reorientate from there. Our camp in this area was just a few tents, which from time to time we shared with crabs, tortoises, lizards, geckos and rats.

My time here was spent making a representative collection of the plants, many of which were quite spectacular. Frequent over much of the area was the screw-pine *Pandanus tectorius* with its stout branches sticking out like arms and bearing dense clusters of large, tough, strap-like leaves and enormous pendulous fruits. This plant is found throughout the Indo-Pacific region, growing near the sea. In contrast *Grewia salicifolia*, a small tree which has clusters of beautiful pink flowers, is found only on Aldabra and the neighbouring island of Cosmoledo. Other trees common in this area were *Ochna ciliata* with clusters of brilliant yellow flowers which emerged in advance of the new leaves, *Thespesia populnea* with large yellow bell-shaped flowers and *Apodytes dimidiata* with small pyramidal inflorescences of tiny white flowers.

Amongst the herbs there were many with colourful flowers which added charm to the vegetation. *Nesogenes prostrata* has tiny mauve flowers; *Cassia aldabrensis* with slender prostrate stems has small yellow or orange flowers – another species which is endemic to Aldabra and the neighbouring island of Assumption.

Apart from trees and herbs there are climbers such as the strange, parasitic *Cassytha filiformis* with its leafless, yellow, cord-like stems; the aloe-like *Lomatophyllum aldabrense*, another endemic with conspicuous spikes of bright orange flowers; *Jasminum elegans* with its long green stems and fragrant white flowers; another leafless plant, *Plumbago aphylla*, which grows in broom-like tufts and bears small white flowers, and *Asparagus umbellulatus*, which adds its feathery stems to the diversity of form found in the mixed scrub.

Giant tortoises in profusion

The remarkable population of giant tortoises is estimated to be in the region of 15,000 animals. This is the largest population in the world and 80 per cent of the animals are located at the eastern end of the atoll, in the mixed scrub forest and along the coastal sand dunes. The reason for such a concentration in this locality is the nature of the terrain.

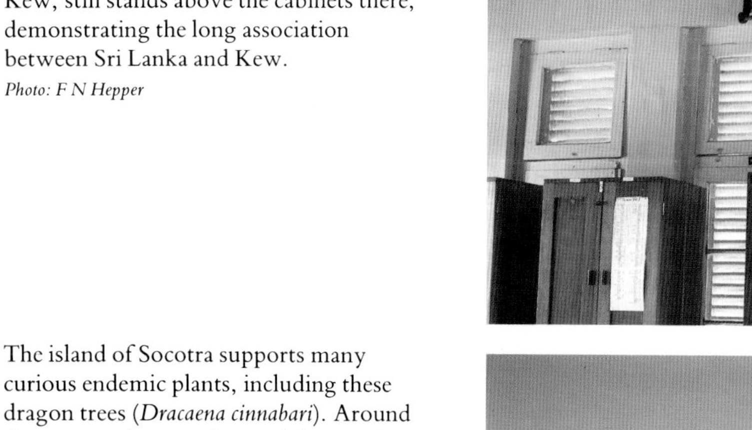

Collaboration with the *Flora of Ceylon* project sponsored by the Smithsonian Institution of Washington enabled several Kew botanists during the 1970s to collect and study plants in the field, as well as in the Herbarium at Peradeniya. A bust of Sir William Hooker, the first director of Kew, still stands above the cabinets there, demonstrating the long association between Sri Lanka and Kew.
Photo: F N Hepper

The island of Socotra supports many curious endemic plants, including these dragon trees (*Dracaena cinnabari*). Around them grow the succulent red-flowered *Aloe perryi* and conical *Adenium obesum*. A Kew expedition to Socotra in 1967 made some very interesting botanical discoveries.
Photo: A Radcliffe-Smith

The Seychelles are now well known for their nature reserves, but the botanical survey there was carried out by C Jeffrey of Kew in 1962. Kew has also recently published S A Robertson's *Checklist of the plants of Seychelles* (1989). The main ridge of Silhouette Island, wide enough for only a single row of trees at its summit, is the home of the beautiful endemic *Glionnetia sericea* and its associate pitcher-plant *Nepenthes pervillei*.
Photo: C Jeffrey

(*above*) A cooperative project between Kew, the Paris Herbarium and the islands of Mauritius and Réunion has been established to prepare the *Flore des Mascareignes*, several parts of which have been published. The spectacular mountains of Réunion are still clothed with forest.
Photo: M J E Coode

(*above left*) One of the famous Seychelles plants, *Medusagyne oppositifolia* in its own family, was thought to be extinct until rediscovered by an expedition to Mahé in 1973.
Photo: M C Whitehead

(*left*) During fieldwork for *Flore des Mascareignes* in Réunion, *Forgesia racemosa* in the family Escalloniaceae was collected. It is one of the endemic genera occurring there.
Photo: M J E Coode

Lomatophyllum aldabrensis is related to the better known aloes of tropical Africa. This genus occurs only in Madagascar and the Mascarene Islands and the species is endemic in Aldabra.
Photo: S A Renvoize

Much of the land surface of Aldabra is fissured or rough coral limestone rock, which is impossible for tortoises to traverse. Over most of the southeastern region the surface is flat slab formations or, on the coast, grassland and sand dunes; both surfaces allow ease of movement in search of food or shelter.

Tortoises are not generally found in *Pemphis* scrub or mangroves. In the mixed scrub the tortoises may browse the lower branches of shrubs and small trees but more important to them is a turf formation composed primarily of small grasses: *Panicum, Eragrostis* and *Sporobolus* and sedges: *Bulbostylis, Fimbristylis* and *Pycreus*. Only these plants can survive the close cropping of the tortoises. Other small herbs such as *Tephrosia, Sida, Asystasia, Euphorbia, Evolvulus* and *Hypoestes* may also occur but they are liable to succumb to intensive grazing unless they are established in a small hole or rock crevice, which protects the base of the plant from damage.

The tortoises usually feed early in the morning and late in the afternoon, when the air is relatively cool. During most of the day the temperature in the open is too high for them to tolerate; they die if they cannot find adequate shade under a tree or shrub. In the mixed scrub shade is seldom a problem; on the coast, however, suitable trees are often widely spaced, consequently the tortoises do not graze more than approximately 300 m (985 ft) from a source of shade. This is the optimum distance they can cover in the limited time of lower temperatures in the morning and late afternoon.

The obvious breeding success of the tortoises in the southeast corner of the atoll and the high density of their population may be their downfall for they are seriously damaging their habitat by the pressure of their numbers. When they gather in groups to shelter under trees from the midday sun, they often dig out the soil around the roots. This ultimately leads to the death of the trees and consequently the loss of vital shade. Also overgrazing or trampling of the grass turf may expose the soil to wind erosion and the intensive browsing of small shrubs and seedlings prevents regeneration.

The extensive development of grassland along the southern and southeastern coast is probably due to several factors. The southeast trade winds, which blow from May to November, undoubtedly carry salt spray far beyond the beach and prevent the spread coastwards of many trees, shrubs and herbs which grow inland in the mixed scrub forest. The three grass species which are most abundant on the coast are salt tolerant: *Sporobolus virginicus*, a pantropical species; *Lepturus repens*, which is widespread in the Indian and Pacific Oceans; and *Sclerodactylon macrostachyum*, which is a coarse tussock grass confined to the western Indian Ocean. In addition to strong winds and salt spray, the tortoises themselves may be responsible for considerable modification of the vegetation in the south of the atoll. Shade trees and shrubs destroyed by tortoise activity are replaced by herbs and grasses, thus increasing the grassland formation even further. In this type of habitat over 30 species of plants are eaten by tortoises.

I made several visits to different localities on the north side of the atoll. Here the vegetation is rather different from that of my previous collecting sites. The island is considerably narrower, there is no grassland and the mixed scrub forest, which forms a correspondingly narrower band than elsewhere on the atoll, extends to the coast. The tortoise population is small in this area and consequently the vegetation has more opportunity to regenerate and is generally much more dense than in the southeast of the atoll. The dense scrub forest in the north of the atoll provides the last refuge in the western Indian Ocean for the white-throated rail. This inquisitive, flightless, brown and white bird, although territorial in its habits is distributed all along the northern part of the atoll. It is quite unafraid of man and will appear out of the bushes in response to any noise or general disturbance.

This rail (*Dryolinnas cuvieri aldabransis*) is one of the few ground-inhabiting birds remaining on the Indian Ocean islands. The famous dodo was exterminated in Mauritius by 1681. On Aldabra this shy rail is confined to the two islands on the north side of the atoll, where the vegetation is particularly dense.
Photo: S A Renvoize

Research rewarded

As a result of this expedition and the following years of activity at the research station, over 150 scientists visited Aldabra, including several from Kew. They have all contributed to a research effort which provides a comprehensive survey of the whole ecosystem of the atoll. The information gathered has been used in planning a programme of management which will ensure that the remarkable plant and animal associations which have evolved on the atoll will continue to thrive. Aldabra will remain a unique environment for future generations to cherish and to gain greater understanding of the world we live in.

6 Plant hunting in the Near East

Plant hunting in the sands of Arabia might seem to be a profitless occupation for Kew botanists, yet there are many parts of the Arabian Peninsula and the Near and Middle East that are well watered or support interesting plants adapted to their dry environment. Kew Herbarium has been publishing the *Flora of Iraq* by C C Townsend and Evan Guest since 1966, and the *Flora of Cyprus* was completed by R D Meikle in two volumes (1977, 1985). Now the *Flora of the Arabian Peninsula* is being researched in conjunction with the Royal Botanic Garden, Edinburgh, and close collaboration with colleagues in Oman and Yemen has resulted in several expeditions by Kew staff to those fascinating countries. This chapter tells of visits to the sands of Oman and the cedar forests of Cyprus, with photographs of other expeditions.

The Wahiba Sands of Oman

T A COPE

I had spent six years working on the *Flora of Pakistan* project and a further six with the proposed *Flora of the Arabian Peninsula*, during which time all attempts at fieldwork had come to nothing – usually for political reasons – so it was with a sense of relief that I accepted an invitation in 1985 to join the Royal Geographical Society's Oman Wahiba Sands Project as botanist. While not an area that would normally have been selected for a Kew expedition, it was an important first step not only in acquiring a first-hand knowledge of Arabian plants, albeit very limited in number, but also in making useful contacts in the country should the opportunity for further fieldwork ever present itself.

A multidisciplinary expedition

The Royal Geographical Society has for many years been associated with large multidisciplinary overseas research projects and of all the terrains visited a true desert was the only important one missing. Thus, it was decided that the 1985–86 project would study a desert environment. The only remaining question was which one? There are many deserts around the world but very few of them lend themselves well to intensive study by a limited number of people in a limited period of time.

It was at the suggestion of the conservation advisor to His Majesty The Sultan of Oman that the so-called Wahiba Sands of Oman were finally selected. The Wahiba Sands have been described, in geologists' parlance, as the perfect 'hand-specimen' of a sand sea. They measure a maximum of 180 km (112 miles) north to south by 80 km (50 miles) east to west, and cover an area of about 9500 square km (3668 square miles) (roughly half the size of Wales). They carry a rich diversity of sand topography, as rich as many much larger deserts; they have a surprisingly extensive vegetation cover in certain areas, even to the point of being described as wooded; they are home to a remarkable diversity of birds, mammals, reptiles, amphibians and many groups of arthropods; and they are inhabited by a large number of human beings of several nomadic tribes.

The effort of the expedition was concentrated in three main disciplines: anthropology, geomorphology and biology. The biological resources team, with which I was associated, was given a fairly wide brief. Three specialists concentrated on cataloguing the animal life of the Sands while a fourth, myself, was there to catalogue the plants. Other members concentrated on the ecology of gazelle populations, water economy of desert plants and genetic resources of the principal woodland tree *Prosopis cineraria*. The project lasted a total of three and a

The high dunes of Wahiba are extremely stable and support a surprisingly rich, if scattered, community of small shrubs interspersed with grasses, sedges and other herbaceous perennials. The infrequent rains bring fresh growth to the shrubs and enrich the vegetation with their carpet of ephemeral species.
Photo: T A Cope

half months from January 1986, though my own input was limited to six weeks, two of which were mostly spent at the Oman Natural History Museum in an effort to become familiar with the known flora.

Preparations

A desert is a hostile environment, not only for its denizens, but also, of course, for visiting scientists. The project was therefore extremely fortunate in being fully supported at all times in the field by the Sultan of Oman's armed forces.

A year before the main research phase of the project began, a small group had made a preliminary survey – the mapping phase – to familiarise themselves with topography, vegetation types and the extent of their cover, possible locations for camps and, of course, access points and safe motor-vehicle routes through the Sands. The sand sea comprises three main topographical units, called High Wahiba, Low Wahiba and Peripheral Wahiba.

High Wahiba includes most of the northern end of the Sands and is characterised by megadunes with a roughly north-south orientation; each dune may be as much as 100 m (329 ft) high and most are many kilometres long; their crests may be 2 km (1.2 miles) apart separated by broad swales. They are relatively stable, remaining largely unaffected by the annual cycle of wind patterns. In contrast, the sands of the Low Wahiba are of lower relief and more varied topography. They also respond much more readily to seasonal and daily changes in wind cycles; many are soft, making travel by motor vehicle treacherous. Peripheral Wahiba comprises those areas beyond the Sands proper, but which are in some way linked dynamically with the whole system; it includes several major wadis and a number of areas where flash-floods fan out. Underlying parts of the Sands, particularly along the coast and in the far southwest, occasionally coming to the surface, are ancient cemented dunes called aeolianite.

The mapping phase of the project was the first serious attempt to

The low dunes of Wahiba are very unstable and subject to considerable seasonal and diurnal movements. The shifting sands are difficult to traverse and support little or no vegetation.
Photo: T A Cope

catalogue the vegetation of the Sands, although numerous naturalists had visited the area since the Second World War, the earliest being Wilfred Thesiger, who passed through in his crossing of the Empty Quarter in 1949. Plant specimens were not again collected from the Sands until 1972, and during the next 15 years only eight other botanists had made worthwhile gatherings. By the end of the mapping phase a total of 22 perennial plant species had been identified in a complex network of ecological associations. Foremost among these associations were two extensive tracts of *Prosopis* woodland, one in the east and one in the west. Apart from a number of fishing villages along the coast, it is in these woodlands that most of the human population resides.

Sand and no rain

The task before me in my four weeks of fieldwork was to collect, identify and document the flora. One very important aspect of this was to make good-quality reference specimens that would be deposited in the national herbarium, and so repeat collections of the species already known were made – at least as far as was practical. No rain had been recorded in the northern end of the Sands for three years and a figure suggested for the southern end was 10 years. This meant that good fresh flowering material was going to be hard to find, a problem exacerbated by a disconcertingly high population of goats and camels. *Acacia tortillis*, a very common tree in the north, was neither flowering nor fruiting, yet *Acacia ehrenbergiana*, in the south, was in full flower; *Prosopis* trees could be found in both flower and fruit. Eventually most of the perennials were successfully recollected, nothing further being possible without good rain to stimulate fresh growth, and attention was turned away from the heartland of the Sands to their margins.

From the botanist's point of view the peripheral regions had much greater potential for species diversity. The wadis in particular, besides being home to all the species found in the Sands, were clearly functioning as reservoirs for a possible extended list of the flora for the Sands if local climatic, soil and other factors allowed the necessary migration. The wadis thus held the gene pool for the entire region and it was important to study them in detail.

Since the main base-camp was sited on the margin of the great Wadi al Batha near Mintirib, it was this wadi that was looked at first. In the immediate vicinity of Mintirib it was nothing more than a broad flat sandy plain little different from the dunes around it, but as we moved eastwards towards the twin towns of Bilad Bani bu Hasan and Bilad Bani bu Ali the dunes to the north gave way to a coarse gravel plain with a quite different vegetation. It was tempting to wander across this plain in the hunt for new plant records, but in the end the survey was limited to within 100 m (329 ft) of the northern edge of the wadi. Narrow channels feeding into the main wadi from the north were the most rewarding, yielding such species as *Lycium shawii*, *Jaubertia aucheri* and *Maerua crassifolia*. The last was never collected as it is an obvious

Schweinfurthia papilionacea, a member of the snapdragon family *Scrophulariaceae*, was seen only once during the Wahiba Sands expedition.
Photo: T A Cope

Where water occurs plants grow. If the water is permanent then cultivated date-palms, other woody plants and weeds form well-developed oasis vegetation, which is also important for other wildlife.
Photo: T A Cope

favourite of camels; it had been repeatedly nibbled until it was reduced to nothing but a dense tangle of short thick woody twigs with few leaves and even fewer flowers. It was, however, successfully photographed. It was on the stony bank of one of these channels that the delightful *Schweinfurthia papilionacea* was found – just once.

Further east, at Bilad Bani bu Ali, there was a further dramatic change in the landscape as the waters of Wadi al Batha rose to the surface and flowed as a permanent stream for just a few hundred metres. Indeed, to reach the Sands from here the wadi had actually to be forded. Water plants and amphibians abounded and dense shrubbery and date gardens lined its banks; but just a few paces beyond, the desert immediately reasserted itself.

Later we left the northern end of the Sands and crossed the interior to the south west, where the underlying aeolianite rose above the desert to form a conspicuous escarpment. The route traversed the western *Prosopis* belt and then crossed an area that had at one time in the relatively recent past been largely under water. Between the dunes there were flat clay pans, some quite extensive, but all bearing a flora of a quite unfamiliar aspect.

Some of the grasses – *Stipagrostis plumosa* in particular – were alive though hardly flourishing, whereas others, such as *Stipagrostis sokotrana* and *Lasiurus scindicus*, were barely recognisable and long dead. Whatever water there had been had left many years ago. The remains of perennial shrubs were unlike those growing further north, so clearly there were species here that have not yet been catalogued. The flora of the two main wadis on this side of the Sands, Wadi Matam and Wadi Andam, was likewise quite unlike that of Wadi al Batha. Chenopods, composites, legumes and grasses were conspicuous, but a tamarisk was by far the most abundant single species.

It is ironic that 10 days after my departure from Oman the long-awaited rains came. Wadi al Batha was transformed from an elongated dry sandy plain to a raging torrent of flood water sweeping down from the Hajjar mountains to the north. Eventually the flood abated, the water trickled away through the sand to its underground aquifer or found its way to the sea and a dramatic change came over the Sands. The shrubs sprang into new life and the tiny seeds of ephemeral plants that had lain buried for years began to germinate. The Bedu moved out with their stock to allow the new grazing lands to develop and eager helpers collected a new range of plants for the Wahiba Sands checklist.

The new additions were rather disappointing in their diversity. This could have been a reflection of the inexperience of some of the collectors, but since these were the same people who had been helping all along that did not seem likely. Perhaps the flora of the Sands was in fact as impoverished as it seemed.

I was given the opportunity to see for myself the following year during the course of a symposium on the project being presented at Sultan Qaboos University near Muscat. The rains that year had come at a good time and in good quantity, and indeed were still falling. A trip

The Sultan of Oman's armed forces provided the use of a helicopter to facilitate plant collecting in remote places. Helicopters are too expensive for general use, but Kew expeditions in various countries have found them a great help in special cases.
Photo: T A Cope

via Mintirib and Bilad Bani bu Ali to the eastern *Prosopis* belt was organised, but at Bilad Banu bu Ali the wadi was in spate and uncrossable; worse, the Bedu track into the Sands beyond was impassable. A diversion via the coastal route to Ashkharah had to be taken and in due course the party arrived at the camp. A disappointing number of specimens were gathered, but they did include a remarkable toadstool, *Montagnea arenaria*, found growing in some of the driest areas.

Camp was finally broken next day and the return trip to Muscat begun. It had been hot at times in January and February in 1986, afternoon temperatures reaching 37°C, but in April the following year it was hotter still. Near Ashkharah the vehicle was stopped so that we could cool off in the sea. It was here that my eye was caught by an unfamiliar succulent herb. I delayed long enough to collect a sample, but in doing so missed my dip in the Indian Ocean. It was worth it though; the plant turned out to be an undescribed species of *Polycarpon*, the only unique plant that the Wahiba Sands can so far claim as their own.

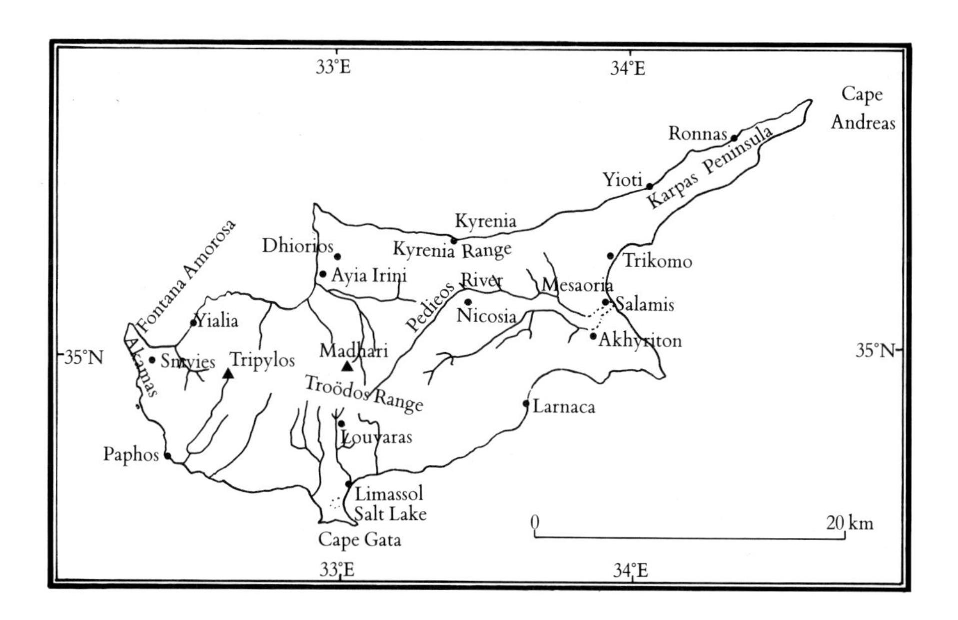

Map 7 Cyprus with places mentioned in 'Springtime in Cyprus', ch. 6.

Springtime in Cyprus

R D MEIKLE

Cyprus is an attractive island for the botanist, especially from March to May, when, after a satisfactorily moist winter, the whole countryside is verdant, the cultivated fields bright with agrestals, and the rocky hillsides clothed with rock-roses, sages and a rich and no less colourful assemblage of shrubs, subshrubs, perennials and annuals.

The topography of the island

The irregular outline of the island, extending in the north east into the attenuate Karpas Peninsula, gives Cyprus a disproportionately long coastline of rocky, shingly and sandy shores, providing ample habitats for a varied maritime flora, with two large salt-lakes sheltering an unusual, and incompletely investigated, association of salt-loving plants, known as halophytes.

Inland, geological diversity has produced two contrasting topographical features which add greatly to botanical interest. In the north a long ridge of limestone rises to 1000 m (3280 ft) in a series of jagged peaks and precipices, forming the Kyrenia Range, the home of many interesting and often endemic plants inhabiting rock crevices. To the west, separated from the Kyrenia Range by the fertile lowlands of the Mesaoria, the more lofty, though less dramatic, igneous Troödos Range, attaining almost 2000 m (6560 ft) in the serpentine-topped Khionistra, harbours a unique flora, with an exceptionally high proportion of endemics, including the famous Cyprus cedar (*Cedrus libani* ssp.

brevifolia), unknown to science until as late as 1878, and virtually confined to one valley in the Paphos Forest, and the Golden Oak (*Quercus alnifolia*), abundant or locally sub-dominant over much of the middle heights. Deeply cut valleys in the western part of the Troödos Range are filled with high forests of *Pinus brutia*, survivals of an aboriginal vegetation which, in prehistoric times, probably covered the greater part of the island.

Earlier collections on Cyprus

From the Middle Ages onward, numerous travellers, halting briefly in Cyprus on their way to the Holy Land, left records of the island and its vegetation; but scientific investigation in the modern sense did not begin until the spring of 1787, when John Sibthorp and John Hawkins, accompanied by the botanical artist Ferdinand Bauer, traversed a limited area of the island, collecting specimens and recording information later published in the *Florae Graecae Prodromus* (1806–16) and in the magnificently illustrated *Flora Graeca* (1806–40). To this basis of information, the 19th-century botanists Theodor Kotschy and Paul Sintenis made extensive additions, subsequently incorporated and considerably augmented in Jens Holmboe's *Studies on the Vegetation of Cyprus* (1914). Thereafter knowledge of the Cyprian flora, and collections of herbarium material, were steadily increased through the industry of numerous botanists, professional and amateur, many of them resident on the island.

By the middle years of this century, it was clear that sufficient data had accumulated for the compilation of a detailed descriptive *Flora*, plans for which had been under consideration at Kew during the period immediately preceding the Second World War, but which were of necessity postponed until peace was restored.

Substantial collections of Cyprian specimens had been acquired by Kew before work began on the *Flora of Cyprus* in 1954. The historic collections of Kotschy and Sintenis were well represented, supplemented by those of A G and M E Lascelles, M Haradjian, A Syngrassides, E W Kennedy, H Lindberg, P H Davis and E C Casey, and these, from 1951 onward, were being extensively augmented by the wide-ranging activities of L P H Merton, then attached to the Department of Agriculture, Nicosia, and in charge of the department's herbarium. Of the more important Cyprian collections only that of Jens Holmboe was unrepresented. It was felt, however, that first-hand knowledge of the island and its vegetation was an essential prerequisite for me as the author of the projected *Flora of Cyprus*. It was clear that while certain areas, around Larnaca, central Troödos and Kyrenia, had been adequately worked, other parts of the island, especially the southwest, the northwest, the foothills of the Troödos Range, and the Karpas Peninsula, were relatively little known, and required further investigation.

One of the surprising plants found in North Yemen is *Primula verticillata*, which occurs in rocky places where water flows throughout the year.
Photo: F N Hepper

The montane plains of North Yemen, at an altitude of some 2400 m, are shallowly terraced and cultivated with Sorghum. Several Kew expeditions have been plant hunting in the Yemen where the broken escarpment to the west of these cool montane plains falls away rapidly to the hot, semi-desert coastal *tihama* with an African-type flora. The new *Flora of the Arabian Peninsula* includes the Yemen plants.
Photo: F N Hepper

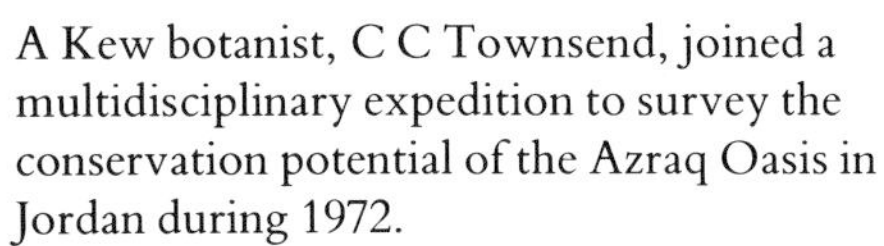

A Kew botanist, C C Townsend, joined a multidisciplinary expedition to survey the conservation potential of the Azraq Oasis in Jordan during 1972.
Photo: C C Townsend

These mountains in the south and west of Anatolia are subjected to winter rainfall followed by warm dry summers, a regime which has favoured the development of plants with swollen underground storage systems such as bulbs and corms, so that there are several hundred species recorded in the area. Many of the newly discovered species are rather rare local endemics and have previously avoided detection because they flower in early spring or in autumn when travelling is more difficult.
Photo: B Mathew

Turkey is very rich in 'bulbous' plants belonging to the iris, lily and amaryllis families. A programme of Kew collecting expeditions and taxonomic research has revealed many species new to science, including this attractive *Iris purpureobractea*.
Photo: B Mathew

Several expeditions to Dhofar in Southern Oman, with a Kew botanist on the one in 1977, studied the plants and animals in the region where this frankincense tree *Boswellia sacra* grows.
Photo: A Radcliffe-Smith

Collecting agricultural weeds in a wheatfield in Kharga Oasis, Egypt, during a Kew visit in cooperation with the University of Cairo in 1985.
Photo: F N Hepper

The foothills of the Troödos Mountains near Layia, west of Pano Lefkara, with *Pinus brutia* and many low shrubs in this type of vegetation known by the Greek word *phrygana*, including *Lithodora hispidula, Cistus* species and *Sarcopoterium spinosum*.
Photo: J J Wood

R D Meikle made three visits to Cyprus in connection with his *Flora of Cyprus* which was prepared at Kew Herbarium and completed in 1985. During his plant-collecting trip in the spring of 1962 he found several plant species not previously recorded from the island.

Fieldwork for the *Flora*

The troubled years preceding independence delayed the possibility of such fieldwork, and it was not until the beginning of March 1962 that I was able to visit the island. In the succeeding 12 weeks, with the guidance, assistance, and hospitality of the Forest Department, I travelled throughout Cyprus, collecting everywhere, but paying particular attention to those areas which were known to be poorly represented in the collections at Kew.

With Nicosia as base, I first paid a fortnight's visit to the southern Akamas, in the extreme southwest of the island. Here I saw the strawberry tree, *Arbutus unedo*, in its only Cyprus station, perhaps significantly in an area where the remnants of Roman settlements are still evident, and collected interesting endemics on the strip of serpentine running from Smyies south to the coast and a surprisingly rich aquatic flora (with *Elatine, Limosella* and Water-crowfoot *Ranunculus* species in shallow, seasonal pools on maritime rocks. The northern part of the Akamas Peninsula – classic sites around the Baths of Aphrodite and Fontana Amorosa, visited by Sibthorp in 1787 – were examined during a stay at Yialia Forest Station.

Then I moved to the sandy area around Dhiorios and Ayia Irini in the driest region of the island. Here I saw in some quantity the leguminous *Argyrolobium uniflorum*, previously known from only one other locality, the attractive but equally rare toadflax *Linaria pelisseriana*, and the rush *Juncus capitatus*, together with other rarities – *Cyclamen graecum, Achillea cretica, Tulipa cypria* – wholly or almost wholly restricted to this region.

After a brief stay at Kyrenia, and a visit to the species-rich Kyparissovouno, the highest peak of the limestone northern range

Cedars Valley in the western Troödos, where the Cyprus cedar, *Cedrus libani* subspecies *brevifolia*, still occurs as a mature, flat-topped tree.
Photo: J J Wood

(with the endemic *Delphinium caseyi* on its summit), I spent nine days during April at Akradhes, in the southern part of the Karpas, with day excursions to almost every part of the peninsula, from Cape Andreas (where the rare crucifer, *Enarthrocarpus arcuatus*, still flourished at the tip of the cape probably in the exact spot where it was first seen by Sintenis in 1880) southwest to Yioti and Ayios Theodhoros. This was a pleasantly varied area, with interesting wet ground in the Ronnas river valley and in the vicinity of Galatia.

Six days at Salamis allowed for extensive collecting in the heavily cultivated eastern Mesaoria, with visits to interesting swampy ground about Akhyritou and the mouth of the Pedieos River. By the middle of April the vernal flora of the lowlands was already passing its best, and I began collecting on higher ground in the splendid pinewoods about Stavros tis Psokas in the heart of Paphos Forest. While there I visited the celebrated cedar forest and saw some interesting plants, including *Arum conophalloides, Papaver postii, Ornithogalum chionophilum* and *Corydalis rutifolia*, on the summit of Tripylos and adjacent slopes.

I moved again, to the highest ground on the island, about Troödos. There were plenty of *Crocus cyprius* and *Paeonia mascula* in flower, also the very rare mouse-tail *Myosurus minimus* and butterwort *Pinguicula crystallina*, but at the beginning of May it was still too early for most of the high-mountain endemics. Indeed it was bitterly cold, with snow falling steadily throughout 6 May, when the lowlands of Cyprus were basking under hot sunshine. It was nice to confirm the occurrence of blue-flowered *Anemone blanda* on the slopes of Madhari. The species was previously thought to be confined to the northern range, and represented, in Cyprus, solely by a white-flowered variant, but the blue-flowered plant had been noted some years earlier in a tourist's

(*above*) There are three species of *Crocus* in Cyprus, all of them endemics. This *C. cyprius* is found in rocky places with *Berberis cretica* (here seen below the summit of Mt Olympus.)
Photo: J J Wood

(*top*) *Corydalis rutifolia* growing in screes on the slopes of Mt Olympus. Although it is widespread in western Asia, in Cyprus it occurs only in two localities.
Photo: J J Wood

posy of wild flowers said to have been picked 'somewhere on Troödos'.

I returned to lower ground and the southern foothills of Troödos at Kakomallis near Louvaras, where I collected the endemic *Alyssum chondrogynum*, abundant on a local outcrop of serpentine. From this centre it was possible to make day trips to Cape Gata, and species-rich fens by the margin of Limassol Salt Lake, as well as isolated peaks at the eastern end of the Troödos Range. I found *Juniperus excelsa*, a juniper tree not previously recorded from Cyprus, on the rocky slopes of Papoutsa; it was later to be found in some quantity, by the Forestry Department, on the summit of Madhari.

The remainder of the visit (12–19 May) was spent in the neighbourhood of Nicosia, sorting and packing specimens, and making arrangements for their despatch to Kew. Time was available for short visits to Trikomo, Kyrenia, Troulli, Larnaca and other places of interest in the vicinity of the capital.

Later visits to Cyprus, in the spring of 1974, and in the autumn of 1981, were brief, and not primarily intended for botanical collecting, though the finding of a new scilla, *S. morrisii*, was an unexpected bonus in 1974, as was the discovery of another novelty, the dandelion *Taraxacum aphrogenes*, in 1984. On both occasions it was possible to visit sites, and to see interesting plants, which had not been seen in the spring of 1962.

Orchis italica growing in damp ground near Polis. This is a representative of the Cyprus orchid flora which is remarkably diverse and numbers some 39 species – they were studied in the field by J J Wood of the Orchid Herbarium at Kew in 1977 in preparation for his account in the *Flora of Cyprus*.
Photo: J J Wood

7 Botanising in mountain Asia

'The wildest dreams of Kew are the facts of Khatmandu': thus wrote Rudyard Kipling in his poem *In the Neolithic Age*, referring to the floral abundance of the Himalaya. Joseph Hooker's expeditions realised some of those dreams when he sent back to Kew seeds of magnificent rhododendrons, magnolias and other garden-worthy plants. In the late 20th century many equally interesting species have been found by Kew expeditions to Afghanistan, Nepal and China. Opportunities had to be taken as they arose, for it happened that as war closed Afghanistan, China opened up following political change which paved the way for close collaboration with Chinese scientists. This chapter relates the experience of two Kew members of staff plant-collecting in these remarkable countries.

To the Wakhan Corridor, Afghanistan

CHRISTOPHER GREY-WILSON

Afghanistan has long had a reputation of being a fierce and hostile country and when I first knew that I was going there in the early 1970s everyone warned about the awful things that were bound to occur. Instead what I found was a very beautiful country and a proud and independent people, curious but hospitable.

The flora of Afghanistan has long interested botanists and plant hunters. Lying in the hub of Asia, sandwiched as it is between the dry bleakness of Central Asia and the wet lushness of India, the flora is very much transitional, reflecting the characteristics of both the European Central Asian element and the Himalayan. Travel in Afghanistan has never been particularly easy but in the late 1950s and the 1960s a series of expeditions ventured into the country, discovering a rich assortment of new or little-known species. The majority of these expeditions

Map 8 Afghanistan and Nepal with place-names mentioned in 'To the Wakhan Corridor, Afghanistan' and 'Floral glories of Nepal Himalaya', ch. 7.

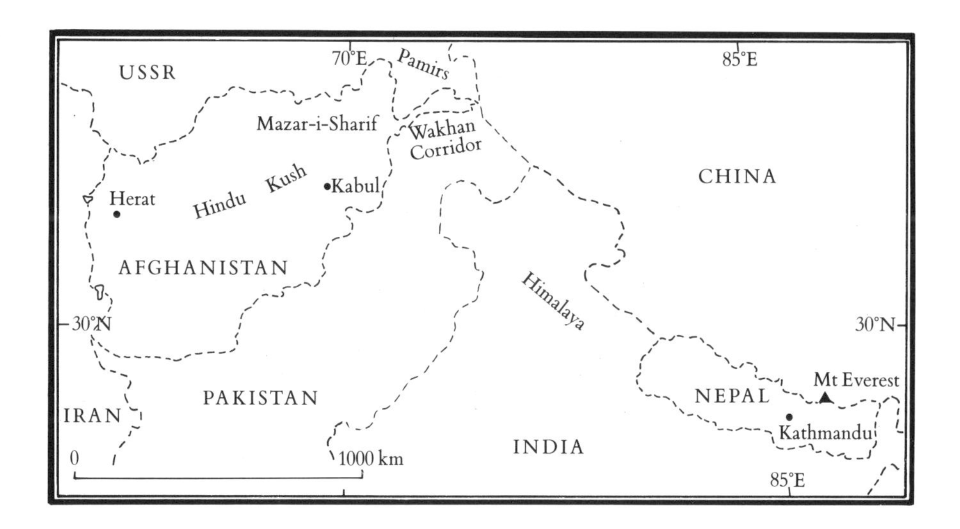

travelled overland from Europe across Turkey and Iran to Herat, or alternatively from the east, from India and Pakistan, over the Khyber Pass to Kabul. Such expeditions often had twofold objectives: firstly, to collect representative samples from different regions for scientific study as dried specimens, and secondly, to collect plants and seeds for botanic gardens and introduction into horticulture. Most prominent amongst those collectors were Per Wendelbo, Ian Hedge, Karl Rechinger and Paul Furse.

Afghanistan, like Iran and the neighbouring parts of the Soviet Union, is renowned particularly for its wealth of 'bulbous' plants. Genera such as *Iris, Tulipa, Eremurus, Fritillaria, Scilla* and *Anemone* are widespread in the region and with numerous species of prime horticultural interest.

Eremurus stenophyllus is a widespread species in Afghanistan, particularly at intermediate altitudes in the Hindu Kush. This region is rich in bulbous and tuberous rooted species that flower in the spring, die down in the early summer to spend the long hot summer period dormant below the ground. *Photo: C Grey-Wilson*

I was invited by Professor T F Hewer to accompany him on a nine-month expedition to Afghanistan and in late January 1971 we headed off from Bristol in a Land-Rover, Kew having graciously given me leave of absence for the period. The journey overland through Europe and western Asia was an unforgettable and rewarding experience, a journey undertaken by many during the 1960s and 1970s, for after all this was the infamous hippie trail to India and Nepal! We had few incidents on the outward journey, but in Turkey, as we climbed over the mountains from the Black Sea coast on to the Anatolian Plateau, the temperatures dropped and we had to endure freezing weather below −20°C. The fear of the vehicle seizing-up forced us to run the engine for 10 minutes or so in each hour throughout the night in such conditions.

Across the border into Afghanistan

The first few weeks we spent travelling and botanising along the Persian Gulf and the Gulf of Oman, but in early April we crossed the border between Iran and Afghanistan and arrived in Herat. During the following six months we explored widely throughout the country, journeying far down dreadful rutted and dusty roads to remote and little-known regions.

Such expeditions are no holiday – the hours are often long and tiring, the terrain dangerous and the threat of infection from local food and water a constant problem. There are often days of travel with few interesting plants to be seen. But the rewards are many – the experience of travel, of seeing new places and peoples and, in Afghanistan, the wonderful sites visited. The glorious views of the great blue lakes of Band-i-Amir, the Buddha cliffs of Bamian, the minaret of Djam, the wonderful blue mosque of Mazar-i-Sharaf and the Great Red Fort near Doäb, laid siege by Genghis Khan, are all memories to cherish forever. Then, of course, there are the rewards of the job: of seeing plants in their wild habitats, of searching areas little-known botanically and of discovering species new to science. The notion of 'discovering species new to science' sounds very romantic but in truth you often do not know whether a species is new until specimens are brought home and carefully analysed, unless of course you have an intimate knowledge of the group concerned.

Afghanistan is often, and erroneously, thought of as a desert country: in fact most of the terrain is of high mountains, fringed by desert only to the south and northwest. The mountains, the Hindu Kush, are a continuum of the Himalaya, rising to over 6700 m (22,000 ft) in the east of the country. They reflect a different flora from much of the Himalaya because they are largely unaffected by the Indian monsoon.

The minaret of Djam, over 600 m (200 ft) tall, of mud-baked bricks and believed to date from 1200 AD, lies in a remote region of the central Hindu Kush, the Hazarajat. It was only discovered by the outside world in the 1950s. C Grey-Wilson, with Professor T F Hewer and the British Ambassador Piers Carter and his wife, journeyed to Djam in July 1971 collecting herbarium specimens and seeds in the limestone gorges surrounding the minaret.
Photo: C Grey-Wilson

Northwards from Kabul

In late May we travelled from Kabul north over the Salang Pass, the road crossing at almost 3700 m (12,000 ft). The green meadows and streams on the north side of the pass are at their best at this time, colourful with early flowering herbs, and dominated by giant fennels (*Ferula* species). A wealth of exciting bulbous plants grow here – including *Iris afghanica* and *I. microglossa*, both in their original (type) localities. Here, in the same vicinity, can be seen *Iris xanthochlora, Ixiolirion montanum, Anemone tschernjaewii* and *Tulipa kolpakowskiana*, amongst a rich medley of other species.

These green areas in the mountains between 2000 and 3000 m (about 7000–10,000 ft) are rich grazing grounds to flocks of sheep and goats. During the summer, countless nomads move from the lowlands into the mountains to seek these green pastures for their animals. The aim of any botanical expedition to such regions is to beat the grazing animals into an area, for once they arrive everything is quickly grazed or nibbled. Amazingly the flora seems to be able to cope with such annual depredations.

To the northeast of the Salang Pass a beautiful and fertile valley, the Andaräb, is fed by waters from the snows and glaciers of the high mountains of Nuristan, densely wooded with pines and cedars. The landscape here is softer, greener and more like that of parts of Kashmir and neighbouring Chitral. Two years previously, in 1969, Professor Hewer had discovered a charming small blue iris of the *Regelia* section and this had been subsequently named *I. heweri*. We returned to collect further material. Close to the town of Banu, the type locality, we came upon several colonies in flower.

Exciting though this was, on a mountain slope within a few hundred yards there were numerous brilliant scarlet tulips, wide-

Travel in Afghanistan is made possible by a network of roads, many only suitable for the toughest vehicles. Few westerners have ever travelled along the Wakhan Corridor where only the first 64 km (40 miles) are motorable. Thereafter sturdy Afghan horses take one across the high mountains, mostly above 3500 m (11,500 ft). Here the Land Rover pauses in the Western Wakhan before negotiating a particularly difficult rock-strewn river-crossing close to the USSR frontier.
Photo: C Grey-Wilson

petalled in the hot sunshine. This later proved to be a new species as well, which I described, naming it *T. banuensis* after the nearby town. Unfortunately the story of this beautiful plant has an unsatisfactory end as bulbs brought back to Britain survived only a few years and the species is no longer in cultivation.

Away from the few main roads in Afghanistan the tracks across the mountains are poor. Usually rutted and with perilous hairpin bends set on steep shoulders, they are frequently blocked by rockfalls so that the best plans can go wrong. Fortunately the previous winter had been dry with unusually little snow. This resulted in few rockfalls and few bridges being washed away by the spring meltwaters. As a consequence we were able to reach nearly all the places we had planned, joining forces on several occasions with the British ambassador, Peers Carter and his wife Joan, who had their own Land-Rover. Joan Carter had collected plant specimens on Kew's behalf during previous years and so it was a useful collaboration.

One of the genera that I had particularly wanted to study was *Dionysia*, a genus closely allied to *Primula*, whose species are almost exclusively adapted to life on shaded or semi-shaded cliffs, often of limestone. Likely habitats abound in Afghanistan and in the last 30 years no less than 10 new species have been discovered there. Most dionysias form dense rounded cushions composed of numerous leaf-rosettes. The flowers are primrose-like, yellow, pink or violet, depending on the species. As we journeyed round Afghanistan we collected a good deal of data on the genus, visiting known localities and finding others.

During August we visited the stupendous gorge of Darrah Zang, crossing the great north-western desert to the town of Maymana. The gorge is one of a complex of gorges to the southeast of Maymana. Only

The genus *Dionysia*, close relative of *Primula*, has a centre of distribution in the mountains of Iran and Afghanistan. There are some 37 species in this region, most being restricted to a single mountain or range. This *Dionysia denticulata* is restricted to a small area south of Bamian, central Afghanistan. Christopher Grey-Wilson made a particular study of the genus while in Afghanistan, discovering a new species, *D. afghanisa* in the northwest of the country.
Photo: C Grey-Wilson

Darrah Zang has been explored to any extent; the others may harbour a host of new and exciting species. Darrah Zang is impressive by any standard with its sheer walls and banded strata affording aspects and niches for a wide range of chasmophytic plants.

This is the finest locality for *Dionysia* yet found. No fewer than five species are to be found there, four being endemic. Besides the widespread yellow-flowering *D. tapetodes*, there were *D. lindbergii* – which curiously grows upside down below overhangs on the ceiling of caverns – *D. microphylla, D. visidula* and *D. afghanica*. The last named was found alone on a large, vertical shaded cliff. There was an extensive colony and, although past flowering, enough remained for an almost complete examination – we knew almost at once that this was a new species. Specimens were duly collected and back in Britain a few seeds extracted. These resulted in a single plant in cultivation which flowered for the first time in 1973. Despite being a very difficult species to grow, *D. afghanica* remains in cultivation to this day. Close by on an adjacent cliff we also found a densely cushioned *Viola* with thickened woody stems and the tiniest pale pink pansy flowers. This also proved to be a new species – later named *Viola maymanica*.

Along the Wakhan Corridor

Undoubtedly the highlight of the expedition was the journey along the Wakhan Corridor, that curious 'finger' of Afghanistan in the northeast, that separates the USSR from the Pakistan and Indian frontiers, reaching east as far as the Sinkiang border of China. The mountains along the corridor, which is essentially the valleys of the Amu Darya (Oxus) and Wakhan rivers, are not the Hindu Kush but the Pamir. The whole area is high, with the upper part of the valley close to the Chinese frontier over 1300 m (14,000 ft) above sea level.

Very few westerners have ever travelled through the Wakhan

The western Wakhan Corridor close to the frontier of the USSR is a land of deep valleys and high passes. Green pastures exist only close to the rivers, and in the higher wide valleys glaciers abound. Most of the Wakhan Corridor lies above 3600 m (12,000 ft) in altitude.
Photo: C Grey-Wilson

Corridor. We travelled in two Land-Rovers to Qala Panj, a small town in the west and thereafter the journey was on horseback. The whole of the Wakhan is bleakly beautiful, with deep valleys and high passes, rugged mountain slopes and scree, with glaciers in the higher valleys. The path is tortuous and treacherous in places with unstable cliff ledges and suspect cantilevered wooden bridges. The eastern half of the corridor has no settlements and the people here are wandering Kirghiz nomads, who make summer encampments of felted yaourts – very similar to the inhabitants of Mongolia.

Collecting plant specimens from horseback is far from easy. The horse either wanders off the moment you dismount (there are few trees in this region for tethering) or eats the specimen you are desperate to collect. Nevertheless we managed to collect over 480 specimens.

The flora here is transitional between the main block of the Pamir to the north and the Karakoram and the western Himalaya to the south. *Clematis hilariae, Myricaria germanica* and *Incarvillea olgae* are common along the river margins. On the high rocky meadows above 4300 m (14,000 ft) most of the plants are ground-hugging alpines, such as *Delphinium brunonianum, Saxifraga komorovii, Gentiana falcata, Acantholimon diapensioides*, forming hard, plate-like cushions which the Kirghiz collect and dry for fuel, *Waldheimia glabra* and *W. nivea*. Little new was collected here, but expeditions are not only about collecting new species, they are primarily concerned with the variety of plants growing in a particular region. In such remote and little-known country every record, even of common and widespread species, is of value.

Today the plight of Afghanistan is known to everyone. From a botanical point of view much still needs to be learnt about the country, and many areas remain unexplored. One day botanists and biologists may be able to return to continue their researches. But first peace must come to the region, and the many thousands of land-mines cleared from the mountainsides.

At high altitudes in the Pamir, as in Tibet (Xizang) and the Himalayas, the yak is the main beast of burden. A rather ponderous animal, the yak is well adapted to such altitudes and more sure-footed on rocky slopes and twisting pathways than a horse. *Photo: C Grey-Wilson*

Floral glories of Nepal Himalaya

A D SCHILLING

(*opposite page*) A window in the subalpine forest to Kantega peak in the Everest National Park, from near Tangboche Monastery. The forests in this region have a rich assortment of species which include *Rhododendron arboreum, Berberis angulosa, Juniperus recurva, Sorbus microphylla* and *Rosa sericea*.
Photo: A D Schilling

Nepal is unique in many ways. Where else in the world is there an altitude variation of 8.85 km (almost 5½ miles)? Where else can you see tropical jungle, stately palms, orchids, coniferous forest, sub-alpine woodland, lush alpine meadows and arid steppe-desert, all within a radius of 160 kms (100 miles) and entirely within the boundaries of one country? Add to these an extraordinarily haunting natural beauty and you have a truly fascinating area in which to carry out plant exploration.

Nepal covers 141,414 sq. kms (54,621 sq. miles) and supports a human population of some eighteen million. Sadly, this fast-growing population (currently the third most rapid in the world) is having detrimental effects on the forests, and the devastation figure for close broad-leaved forest is in excess of 800 sq. kms (300 sq. miles) per annum from a total area of 16,100 sq. kms (6200 sq. miles). The country contains an estimated 6500 species of flowering plants (315 of which are endemic), 30 species of gymnosperms and approximately 450 species of ferns. Sino–Japanese botanical species dominate the eastern and central parts of the country, West Himalayan and Mediterranean elements affect the west, whereas Central Asian plants intrude into the arid rain-shadowed regions of the north and Indo-Gangetic factors make their mark on the southern foothills and plains of the Terai.

The story of botanical exploration in Nepal is a protracted and fascinating one, spanning almost two centuries. Many of our familiar garden plants originally came from the hills and valleys of Nepal – brought back during the 18th and 19th centuries by such intrepid pioneers as F Buchanan-Hamilton, N Wallich, Jospeh Hooker, B H Hodgson and E Gardner. In the latter part of the 19th and the first half of this century, Nepal became politically insulated against western intrusion and, with the exception of limited visits by I Burkill and F M Bailey, it was 'a forbidden land', although two Nepalese travellers (Dhwoj and Sharma) continued to send material to British gardens via royal connections. Then, after the democratic coup of 1949, Nepal's isolationism relaxed and the 'doors' reopened for exploration. First Oleg Polunin, then many others, especially Adam Stainton, studied the flora of Nepal and in consequence it is now one of the best known in the Himalaya. However, it still keeps some of its secrets. I have been fortunate enough to unlock some of them and have experienced at first hand the pleasures such surprises bring.

I became obsessed by mountains even before I became involved in plants. Even in my earliest youth I was drawn to the high hills of Britain and it was there that the two interests first came together. In my twenties the higher mountains of Europe brought me further fulfilment, but it was always my ambition to tread the wild paths of the

Rhododendron arboreum woodland of the Dharchy Forest, Gurkha Himal at 2750 m (9000 ft). This region of central Nepal has been much depleted by deforestation.
Photo: A D Schilling

The Everest National Park. Climbing along the edge of the Khumbu Glacier near Lobuche. At this elevation – about 4000 m (13,000 ft) – in the early autumn the snow is already deep, blanketing a rich assortment of low shrubs and herbs.
Photo: A D Schilling

Himalaya. So in 1965, when Sir George Taylor (who was then director of Kew) suggested that I apply for the position of adviser to the Royal Government of Nepal in the development of a national botanical garden, I needed very little prompting.

The job was all that I could have hoped for but it also gave me the opportunity to get out into the vastness of the Himalaya to explore for plants. So began my continuing relationship with Nepal. The time initially spent in Nepal coupled with that of 16 subsequent visits amounts to three whole years of my life, years which have brought me much pleasure and excitement as well as occasional hardship.

In spite of what the uninitiated might think, plant collecting is not 'a bed of roses'. The so-called 'romance' of plant hunting exists mainly in the imagination of the armchair explorer and largely retrospectively, if at all, in the mind of the actual hunter. Leeches, fatigue, illness, discomfort and monotonous diet are all too frequent. In reality, new and exciting plants are not discovered on every bend of the trail, nor on the crest of every other pass; they are often hard won or stumbled upon when you least expect it. Sometimes it happens years after collection, when a plant which has been introduced and grown in cultivation for some time under one name turns out to be something new to science.

Finding plants in Nepal

In 1975 I collected seeds from a herbaceous spurge in the Dudh Kosi valley on the long, hot trail to Mount Everest. It turned out to be a very attractive and long-flowering plant and was labelled at Kew and elsewhere as '*Euphorbia sikkimensis* – good form'. Almost 10 years later, and after numerous promptings from learned colleagues, material was re-submitted to the botanists and the plant was recognised as a new species. Alan Radcliffe-Smith, of Kew Herbarium, generously named it in my honour and the gardening world now has to live with the name *Euphorbia schillingii*. Apparently, the main differences between my plant and the closely-related *E. longifolia* are that mine has shorter, broader leaves and produces only partially warty fruits. What a way to be recognised!

Over the years I have been fortunate to glean several other first-rate garden-worthy plants from Nepal. In 1965, while engaged in a government botanical survey of the Langtang Valley, I collected a seedling cherry tree with attractively peeling bark. It was rather like the long-cultivated Tibetan species *Prunus serrula*, but with a distinctively hairy leaf and a much deeper brown shiny bark; it has proved to be faster growing and in consequence may prove to be of commercial value as a good tree for the smaller garden. Its name is *Prunus himalaica*. On the same expedition we also discovered a hitherto unknown alpine meadow grass – *Poa langtangensis* – but this has excited only the specialists.

Euphorbia schillingii, a recently described species discovered by A D Schilling in the Dudh Kosi valley on the trail to the Mount Everest region. This attractive species is now well established at Kew and other western gardens.
Photo: A D Schilling

Ginger lilies have always fascinated me and I have discovered and introduced two new ones from Nepal. The first, which was growing in the forest south of Everest, was a clonally distinct form of *Hedychium densiflorum* and in 1982 it received an Award of Merit from the Royal Horticultural Society. It has been named 'Stephen' after my son, who was born in Kathmandu shortly after that particular expedition ended in the monsoon of 1966.

Several years later I took seeds from a withering plant of *Hedychium coccineum* which was growing on the eastern rim of the Kathmandu Valley. As luck would have it – casual collections are always a gamble – these produced a form of exceptional hardiness and startling flower colour. In 1984 it received a First Class Certificate from the Royal Horticultural Society and was named 'Tara' in honour of my daughter, whose name is the Nepalese for 'star'.

Over the years other good plant discoveries or reintroductions have come my way as I trekked throughout the central and eastern parts of the country. These include *Cotoneaster cavei, Rodgersia nepalensis, Salvia castanea, Populus glauca, Corylus ferox, Skimmia laureola* var. *multinervia* (10.5 m (34 ft) tall and black-fruited), *Elaeagnus tricholepis, Euonymus frigidus* var. *elongatus,* and the beautiful *Primula boothii* var. *alba*. Other species I have brought back to Kew and Wakehurst Place have yet to be properly identified, and there are also what are hopefully hardy and improved colour forms of better known plants such as the sweet-scented, winter-flowering deciduous *Daphne bholua* var. *glacialis*.

During my visits to Nepal there have inevitably been times when morale has been low. When struck down in camp with illness or perhaps recovering from the trauma of crossing a dangerous bridge or landslip, I confess that I have sometimes yearned for the safety and comforts of home. But not for long. The quest for plants is an insatiable and compulsive instinct and I am soon musing again on what might be around the next turn in the trail.

Map 9 South western China with places mentioned in 'China's Yunnan Mountains', ch. 7.

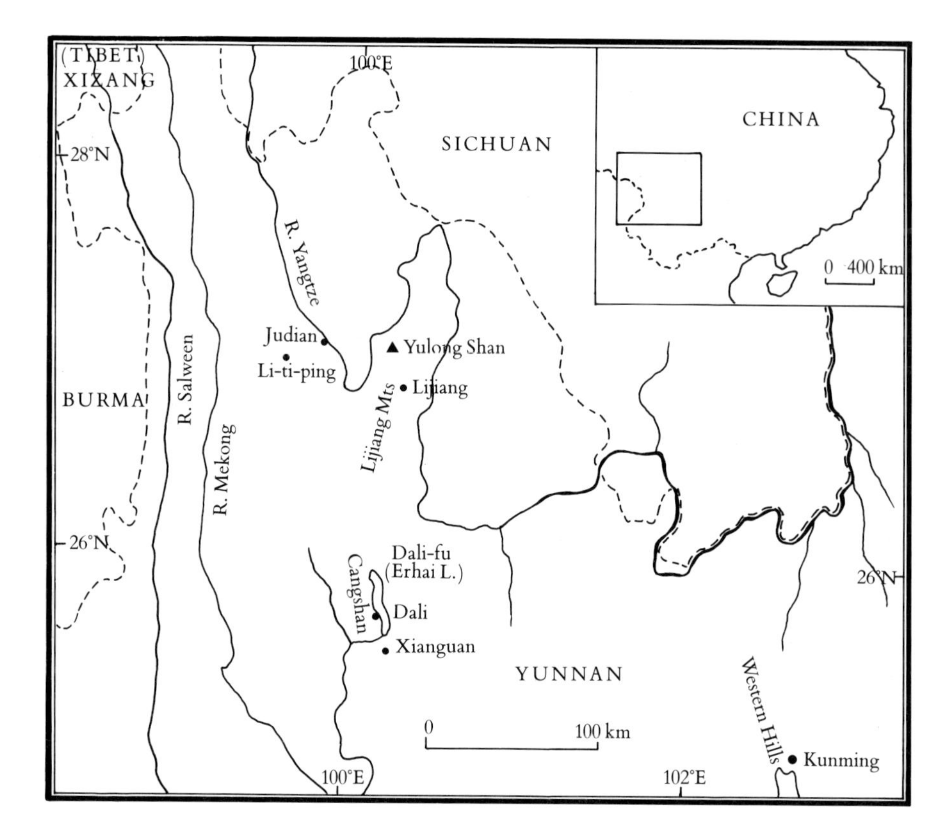

China's Yunnan mountains

CHRISTOPHER GREY-WILSON & A D SCHILLING

The golden age of plant hunting was without doubt the latter half of the 19th century and the first half of the 20th century. One of the chief goals of plant explorers was China. As early as the 18th century Europeans had realised that in China there was an extraordinary and varied richness of plants, and it was quickly decided that there were many species that would prove to be hardy in gardens in Europe and North America.

China, though, is a huge and diverse country. The south enjoys a tropical or subtropical climate, the north a bleak boreal one. As a result the flora is not only extremely varied from place to place but very rich in species. Many of the plant collectors were drawn to the more temperate regions – to Yunnan, Sichuan, Kansu and Xizang (Tibet) – in their search for likely species to enhance western gardens. They came from a number of countries, but primarily from Britain, France and Ireland, and included some of the most famous names in plant collecting – Abbé Farges, Abbé Delavay, Reginald Farrer, George Forrest, Augustine Henry, Frank Kingdon Ward, Frank Ludlow and George Sherriff, Joseph Rock and Ernest Henry Wilson.

Many of these plant hunters were explorers in the true sense of the word, greatly adding to our knowledge of the geography of the regions in which they travelled. They introduced innumerable fine plants to our gardens and many of these, which today we take for granted, are reminders of the diligence and enterprise of this remarkable breed of men. To quote more than a very few examples would simply create a tedious list, but some of the finest must be mentioned as their very names have immortalised the great men who were associated with their discovery: *Catalpa fargesii, Abies delavayi, Cotoneaster franchetii, Picea farreri, Hypericum forrestii, Parthenocissus henryi, Rhododendron wardii, Primula sherriffii, Rhododendron ludlowii, Malus rockii* and *Magnolia wilsonii* are but single examples selected from the Chinese 'botanical encyclopedia'.

After the Chinese revolution in 1911 European plant-hunting activities diminished until by 1950 such exploration had more or less ceased altogether. Frustrated by the closure of China, western plant hunters and botanists turned their attention to other regions, particularly to the Middle East, to Turkey, Iran, Afghanistan, the central Himalaya and to temperate regions of South America. But always the draw of China remained – the thought that in that vast region there were many more exciting species and forms yet to be brought into cultivation.

In the 1970s, as the effects of the Cultural Revolution began to recede, so China opened its 'doors' once more to the West. This, however, was a different China, a China no longer prepared to be exploited by the First World, determined to seek its own destiny. Plant hunters and botanists began to drift back into China, but they were severely restricted as to where they could go and what they could collect. Unfortunately, a number of expeditions chose to ignore the

Close to the Yangtse River rise the western flanks of the Lijiang (Lichiang) Range or the Jade Dragon Snow Mountains, in north-west Yunnan – a high, primarily limestone, range rising to over 6100 m (20,000 ft). This spectacular, although rather inaccessible, region was the chief goal of the 1987 Sino-British Lichiang Expedition, with members from both Kew and Edinburgh botanic gardens and the Institute of Botany in Kunming, capital of Yunnan province.
Photo: A D Schilling

guidelines of the Chinese authorities and abused the rules of collecting. This has meant that severe restrictions have now been placed on the collecting of live plants and seeds. Expeditions have to work closely with the appropriate authority – the cooperative venture in which both Chinese and foreigners benefit equally proved to be the way ahead. The Chinese themselves have embarked upon a huge scheme to write a flora of China, the *Flora Sinicae*, as well as numerous regional floras.

The chance for Kew botanists to visit China came in 1987, when a joint British–Chinese expedition to the Jade Dragon Snow Mountains (Lichiang Mountains, Yunnan), with their highest peak Yulong Shan, was proposed. So we teamed up with two from Edinburgh Botanic Garden, David Chamberlain and Ron McBeath, and in early July flew to Kunming, the capital of Yunnan Province.

Towards the Jade Dragon Snow Mountains

In Kunming we met our three Chinese colleagues at the Institute of Botany, Professor Li-yun-chang, Yuan-lui-kun and Hsu Ting-zhi. Although our stay in Kunming was short, we took the opportunity to visit the fabled Western Hills, where we obtained a tantalising introduction to the botanical riches of Yunnan. Here we had our first sightings of *Rhododendron microphyton, Clematis chrysocoma* and *Vaccinium fragile* in flower. The immature seed pods of *Cercis yunnanensis* glinted memorably close to a cliff-top temple and stately pencil-like specimens of *Cupressus duclouxiana* grasped the craggy ledges high above the great lake of Kunming. After two days' preparation we headed off along the road northwest from the city. The personnel were divided amongst two vehicles, together with collecting equipment, luggage and a prodigious amount of bottled beer – a good basis for an expedition! Our driver drove 'adventurously', mostly on the wrong side of the road and intent on challenging every oncoming lorry.

That first night we arrived in Xiaguan on the southern shore of Dali-fu (Erhai Lake). The following day we continued northwards past the historic town of Dali with its splendid gateways and wonderful white pagodas. The landscape became increasingly interesting: the deforested slopes we had observed further south now replaced by patches of woodland and neat villages that can scarcely have changed for hundreds of years.

The road eventually wound over a high pass where white roses, *R. longicuspis*, small trees of the cream-bracted *Cornus capitata*, with tiered branches, and pink and white forms of the sweet-scented *Rhododendron decorum* flourished. *Rhododendron yunnanense* also abounded close to the top of the pass, where we looked north to Lichiang (Lijiang) and the snowy peaks of the Jade Dragon Snow Mountains. This superb massif, which rises to 6100 m (20,000 ft), is the same age as the Himalaya to the west and is botanically a very rich area. The first systematic exploration was made by the French missionary collector Abbé Delavay in the 1890s, but it was left to the Scotsman George Forrest to introduce many of the plants of the region into cultivation. He returned on

Northwest Yunnan is rich in mixed evergreen and deciduous forest. Here rhododendrons abound, such as the sweetly-scented *R. yunnanense*, seen here growing at Gang-ho-ba north of Lichiang (Lijiang). Early plant explorers collected seed of many rhododendrons and numerous other plants in this region, many of which have proved hardy in British gardens. Today, the political situation in China is such that western botanists can once again visit these remote parts of the country in search of horticulturally and botanically interesting species.
Photo: A D Schilling

repeated expeditions to Yunnan at the beginning of the century and his detailed knowledge of the Lichiang (Lijiang) mountains and his use and training of native collectors has become a legend.

The botanical riches of the Lichiang (Lijiang) region

But why, the reader may ask, had *we* come here? Every expedition has an aim and ours was to collect representative herbarium specimens and seeds. Despite the work of earlier collectors, many of the plants of the region are still little known and good modern collections are invaluable in giving a clear picture of the species, their range and variability. Seed is equally valuable for there is a good chance of introducing further material of species rare or little known in cultivation or indeed not yet introduced. It is also possible to find clones which are horticulturally superior to those already in cultivation.

Over the next five weeks we made repeated excursions into the mountains. Unfortunately, we were not able to camp as we would have wished, due to local bureaucratic difficulties. This meant that we had to journey daily from our hotel in Lichiang, at about 2700 m (9000 ft) altitude, along the long dusty road north towards the eastern flanks of the mountains and then climb up the slopes, this on several occasions requiring a climb of some 2000 m (7,000 ft) and back in the same day – a desperately exhausting and altogether unsatisfactory method of carrying out plant hunting. This daily effort of ascent and descent collectively totalled 36,000 m (120,000 ft) – equivalent to four times the height of Mount Everest! Added to this, the range is inaccessible for the most part, the low wooded flanking hills ended abruptly in soaring rock walls, steep-sided gullies and difficult ridges. High above, the snow peaks and tumbling glaciers beckon if only one could find the way up.

Androsace rigida photographed at Lichiang (Lijiang), Gang-ho-ba, inhabits stabilised moraines and open woodland. This pretty plant has proved very difficult to cultivate, but seed collected by the expedition may produce easier forms to grow.
Photo: C Grey-Wilson

Roscoea humeana, a spectacular ginger, inhabits rock outcrops north of Lichiang (Lijiang), its flowers appearing before its leaves. This is one of a large number of familiar garden-plants seen by the expedition in the wild. Earlier in the present century the same species was observed in the region by both George Forrest and Frank Kingdon Ward, two of the greatest plant-hunters to visit China.
Photo: C Grey-Wilson

This was exciting country and each day was punctuated by cries as new discoveries were made, but it was also enthralling to see so many familiar garden plants in their wild habitat. The reader must picture woodland rich in trees like *Betula utilis* var. *prattii, Acer forrestii, Populus szechuanica, Tilia likiangensis*, a medley of colourful rhododendrons, *Abies delavayi, Picea lichiangensis*, bushes festooned with *Clematis montana* and an undergrowth rich in herbaceous plants, especially species of *Cypripedium, Roscoea, Euphorbia griffithii, Iris pseudorossii* and many more besides. Wet flushes in clearings provided a habitat often rich in bog-loving primulas and the extraordinary, almost black-flowered, *Iris chrysographes*.

The alpine plants were hard won as we trudged wearily up steep slopes, bypassing treacherous cliffs and scrambling over coarse knee-cutting screes, but the rewards were many. Some of the loveliest of alpines inhabit these mountains: *Paraquilegia microphylla, Solmslaubachia pulcherrima*, a cushion crucifer with stunning ice-blue flowers, the deep reddish-pink *Primula dryadifolia* that has defied the finest alpine cultivators, the charming white *Androsace delavayi* that occasionally produces a pink form and perhaps the smallest of all figworts, *Scrophularia chasmophila*.

There were rare plants, unknown, or little known, in cultivation, which we were desperate to introduce – *Betula calcicola*, the grey-mauve flowered *Indigofera pendula*, the yellow flowered *Stellera chamaejasme, Piptanthus tomentosus*, the quite breathtaking yellow *Incarvillea lutea* and the exquisite little blue poppy *Meconopsis delavayi*. George Forrest had introduced several of these years before but they had failed to succeed in cultivation and only time will tell if we and later expeditions have any more success.

Visiting China is exciting but often frustrating – the pace of life is slow and it is often impossible to speed things along or to take a short cut. Above all, China is a culinary adventure, a delight to those who enjoy a varied diet but full of hazards for the unwary – steamed chicken and caterpillars or duck stomach soup are not everyone's favourite dishes! We found the steamed dough-buns wholly inedible and decided that apart from fish bait for piranhas the best thing to do with them was to hurl them at one another – but only out in the open, of course!

Christopher Grey-Wilson of Kew and Ron McBeth, Curator of the Alpine Department of Edinburgh, collecting seeds in the Lijiang (Lichiang) Mountains during the joint Kew-Edinburgh botanic gardens expedition to north-western Yunnan in 1987.
Photo: A D Schilling

The Mekong – Yangtse Divide

The three great rivers of Western China, the Salween, Mekong and Yangtse, form close watersheds to the west of Lichiang near to the border of Burma and Xizang (Tibet). These remarkable steep and narrow divides are extremely difficult to penetrate but past plant-hunters found them a rich field for exploration. The unexpected chance came for us to visit briefly Li-ti-ping on the Mekong–Yangtse Divide – an opportunity we took up enthusiastically. The long drive from Lichiang took us west round the southern flanks of the Yulong Shan to the Yangtse Bend, where the great river turns in a meander of 270 degrees to head not south like the Salween and Mekong, but east right across China. En route we saw the strange *Photinia prionophylla* with holly-like leaves, so different from other species we know. We then followed the Yangtse along a dry and dusty rough road towards Judian and Li-ti-ping, an eight-hour journey. By now the monsoon should have been in full spate but it scarcely rained during our entire stay in China! In consequence, we were not unduly troubled by leeches but the dry conditions probably caused us to miss seeing that gem of north-western Yunnan – *Primula vialii*.

The woodlands of the region are rich and both Kingdon Ward and Forrest had collected here. Since their day, however, forest destruction has greatly accelerated and at times we were greatly shocked by the felling all around us, dismayed by the wastage and saddened to see the constant stream of lorries removing timber. Forest burning is another major problem and we saw plenty of evidence of such practices.

Yet much still remains to be seen and enjoyed. Here our greatest memory was the forests and the majestic and varied assortment of trees and the ridge-top glades surrounded by thickets of rhododendrons in profusion – white *R. phaeochrysum* and *R. fictolacteum*, pink *R. beesianum, R. leptothrium, R. dasycladum* and *R. oreotrephes*, yellow *R. wardii* and *R. mekongensis* and many more besides. In all, we saw some 40 rhododendron species during the entire expedition.

In addition to the rhododendron an incredible array of other woody species occurred in this especially rich corner of Yunnan. In the warmer lower woodland we saw that strange evergreen relative of the oak, *Lithocarpus pachyphylla*, the neat-leaved holly, *Ilex yunnanensis*, as well as *Cephalotaxus fortunei*. Later, as we gained height, we passed into cool temperate forest boasting superb 28 m (90 ft) tall specimens of *Betula utilis*, as well as fine examples of *Larix potaninii, Tsuga yunnanensis, Acer forrestii*, and *A. giraldii*. Beneath the forests, and at their edges, were many interesting smaller plants, including *Deutzia monbeigii, Osmunda*

Paeonia lutea, widely grown in western gardens, is native to Yunnan and Southeast Tibet (Xizang). The Sino-British Lichiang Expeditions came upon it in flower in June 1987 in the rich forests of the Mekong-Yangtse Divide near Litiping. The wild form seen here differs from those in cultivation by having a prominent orange blotch at the base of each petal.
Photo: C Grey-Wilson

claytoniana, Enkianthus chinensis, Sambucus megalophyllus and a particularly good form of *Paeonia lutea* with rich red markings at the base of the petals – the list was seemingly unending.

The curious arum-like arisaemas were everywhere in the woodland and scrub. They are a group that fascinate some but repel others. Their snake-like spathes and tongue-like spadices are infinitely variable from one species to another. The commonest two in the Li-ti-ping region were *Arisaema handelii*, with green and white spathes, and *A. wilsonii*, with spathes striped with purplish-black and white. The genus is of particular interest as it is being studied at Kew for a future monograph so we were keen to study and photograph them as thoroughly as possible.

There were other noteworthy plants, such as the little-known marshland *Lilium souliei* with its dusky mahogany bells, like a huge *Fritillaria*, and the exquisite yellow-flowered Lampshade Poppy, *Meconopsis integrifolia*, each flower some 14–15 cm (5½–6 in) across. Where the grassland was well established, drifts of yellow *Stellera chamaejasme* clothed the slopes, a remarkable plant with pingpong-sized balls of small daphne-like flowers. This species is the Tibetan paper plant, widely used for making paper, its thin stems tough and fibrous.

China is an amazing experience for any westerner. For the biologist it is a haven, for the plant hunter simply paradise. Yet many of the localities visited in the past by plant hunters have since been destroyed. This destruction continues unabated and the future for many areas is in doubt. We must at least be thankful that so many fine Chinese plants are already well established in our gardens. And we should spare a thought for those earlier plant hunters who often endured great privations as well as many dangers in their search for new introductions to cultivation.

8 Botanical exploration in Papua New Guinea

MARTIN J S SANDS

The large tropical island of New Guinea is neatly divided into the western half – Irian Jaya which is part of Indonesia – and the eastern portion which, together with other islands including New Ireland, is independent Papua New Guinea. Exploration is hindered by the rugged terrain and thick forest cover. Botanically, Papua New Guinea is remarkably rich in plant species and to this day much remains to be discovered. Only in the last two decades has Kew become directly involved in fieldwork there and this chapter records primarily some of the events of an expedition in 1975.

Suddenly the small government trawler *Eros* glided into view from behind the southern headland of the bay as we stood watching on the palm-fringed beach at Natkumlagia. Despite our idyllic Pacific island setting, what a welcome sight that ship was after almost a week of anxious waiting, and after sighting only occasional vessels far out to sea, it seemed immense. For days, with Mark Coode, then of the Forestry Division of Botany in Lae, I had been on stand-by to depart. We had had no idea how long it might take for Clive and Alan – still with an infected leg wound – to make the journey by coastal trail and canoe to the radio at Lambon, a small island off the southern end of New Ireland. Even then we were not to know how soon the boat would respond to a signal diverting it from our original rendezvous on the western side of the island. So uncertain had we become as to when the pick-up boat would arrive that, with the next stage in my collecting programme on Manus Island scheduled to begin very soon, we had resorted to signalling any passing vessel, using only a piece of broken mirror and the sun!

Thus, on 4 March 1970, the New Ireland phase of a memorable botanical expedition to Papua New Guinea came to an end. I stood with Mark on the deck of the *Eros*, waving to diminishing figures on the shore, the details of the village finally vanishing in the heat haze as

Map 10 Papua New Guinea, including associated islands such as New Ireland, with place-names mentioned in ch. 8.

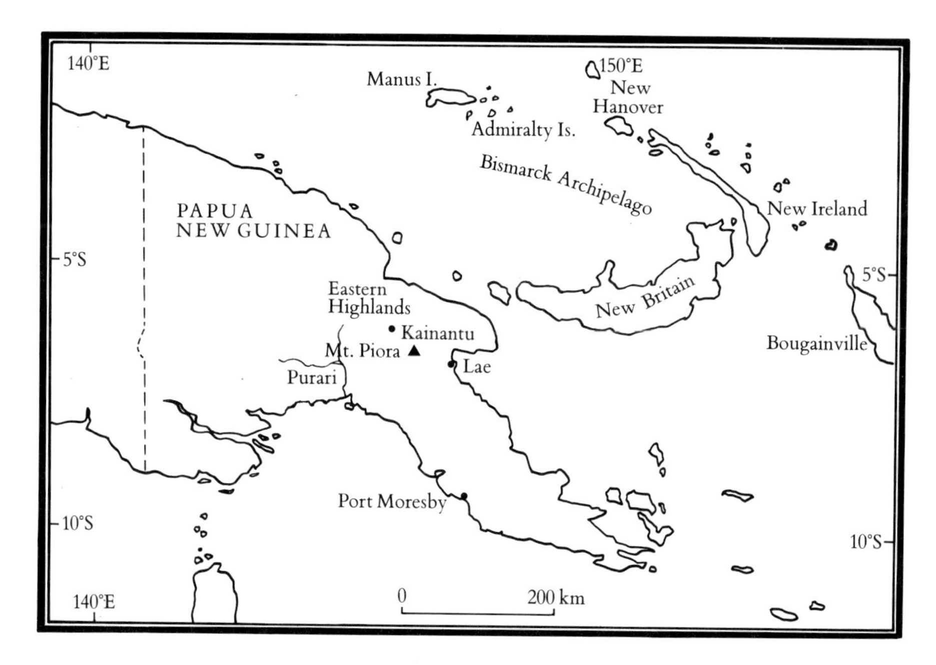

we sailed out to sea and southwards to Cape St George. For nearly six weeks, with our expedition leader, Clive Jermy, and moss expert, Alan Eddy, of the Natural History Museum (London), we had penetrated deep into the island's interior a long way to the north of Natkumlagia, establishing a camp on the banks of the Danfu River some 8 km (5 miles) inland from the east coast.

Early on the morning of 22 January that year, we had landed our cargo on the beach at Muliama plantation from another government trawler, the *Andewa*, soon establishing a temporary camp where the coastal track crosses the Danfu a little to the south of Manga Mission. A trail was cut following the valley and before long the main camp upstream was operational. Trekking inland, through forest in which large trees of *Terminalia brassii* were common, took a number of hours and involved crossing the river several times, often wading waist-deep in the fast-flowing current.

(*opposite page*) Lowland rainforest on the eastern side of southern New Ireland. Near to water, where there may be a break in the forest canopy, a dense undergrowth of shrubby and herbaceous species can occur. In a ravine, during the 1975 Kew expedition, the fern *Christensenia aesculifolia* was discovered, known also in Indonesia but, within Papua New Guinea, only previously recorded from Bougainville. To reach the mountains of the interior involved cutting a trail through rainforest and sometimes wading waist-deep in fast-flowing rivers.
Photo: M J S Sands

Those weeks in the mountains were exciting and sometimes dangerous. Frequent torrential rain and flash floods from storms further inland often made the valley trail to the coast impassable. On such occasions the brown, foaming waters flooded part of our camp site and, with a deafening roar, sent huge boulders rumbling inexorably on their way to the sea. If this were not exciting enough, giant trees crashing about us, when the Danfu valley was caught in the edge of a cyclone, were frightening indeed and the only course of action, as several huge branches fell on the camp, was to stand between the buttress flanges of a large *Pometia pinnata* and pray that it too would not be a victim of the violent winds. On another occasion, a fig fruit, falling in the night from a plant press onto the drying stove below, caused a fire in the polythene work-shelter. In consequence nearly 100 specimens were lost as well as a range of garments drying after a particularly wet day in the forest. Even as we were leaving the valley for the last time, I fell on deeply

weathered limestone, severely cutting one of my fingers with a bushknife and for several hours had to hold the tissues together before a proper dressing could be applied. Much more seriously, Alan suffered greatly when he stumbled on a sharply cut sapling, driving into his calf a splinter several inches long which could only be extracted by strong New Guinean teeth.

We left the forest on 20 February and headed south along the coast with specimens and equipment borne by several changes of carriers as we moved from one small village to another. Our intention had been to cross the island by a great dividing valley thought to follow a fault line and occupied by the Kamdaru River flowing northwestwards and the Weitin River, which flows southeastwards, reaching the Pacific a few kilometres to the north of Natkumlagia. However, after three days of walking, Alan's wound and associated fever were giving cause for concern and, furthermore, it seemed that the only person who knew of the route over the watershed was an old man, in a village near the Weitin mouth, who was no longer fit enough to make the arduous journey. Accordingly, with our scheduled time in New Ireland running out, the island crossing was abandoned and we continued to Natkumlagia.

New Ireland's location and history

New Ireland is a narrow island more than 350 km (220 miles) long in the Bismarck archipelago and, with its smaller, satellite islands, forms the most northeasterly outpost of Papua New Guinea. Lying to the north and east of New Britain and northwest of Bougainville, it occupies an area of nearly 10,000 sq. km (3900 sq. miles) but even its widest, southern portion is only about 50 km (30 miles) across.

While, both botanically and in other respects, many parts of the island remain little known and unexplored even today, New Ireland was in fact one of the earliest parts of Papua New Guinea to be discovered by Europeans. The Dutchmen WC Schouten and J Le Maire sailed past the north of the island in 1616 and then, in 1700, at the eastern end of New Britain, the buccaneer and explorer William Dampier, named St George's Bay. It was 67 years later that the Englishman Philip Carteret recognised this bay as being, in reality, a channel separating New Britain from land to the east which he called New Ireland. More than a century later, missionaries first visited the island and, in 1908, Karl Sapper, a geographer on a German expedition, explored in the southeast. In that same year, Gerhard Peekel began several decades of Roman Catholic missionary service stationed in New Ireland, where he collected plant specimens from time to time, several genera subsequently being named in his honour.

In the late 19th and early 20th centuries, as Neu Mecklenburg, New Ireland was one of the most profitable parts of German New Guinea. A rigorous administration, under the leadership of Baron Boluminski, established important copra plantations in the northeast of the island and cut the first reasonably good roads in all the territory which was subsequently to become united as Papua New Guinea.

Much of the population of about 60,000, mostly of Melanesian stock, lives in the narrow, northern two-thirds of the island with the majority in and around the small towns of Kavieng and Namatanai. In the south, the people nowadays live almost exclusively in very small communities by the sea, leaving the rugged and forested interior uninhabited. Our attention had been particularly drawn to these botanically unexplored forests and the shortage of people for hire as carriers in the sparse coastal population was one of several factors making expedition plans difficult to bring to fruition.

Planning to return

The intention, during that first visit to New Ireland in 1970, had been to penetrate the unknown interior in order to collect botanically from as diverse a range of habitats and altitudes as possible. However, although our collecting programme in the valley was, in itself, very successful (over 3000 specimens), lack of time and the waterfalls and narrow gorges of the Danfu prevented the party from exploring any of the forest above about 1100 m (3600 ft). The much higher mountain ridges thus remained unknown and presented a challenge which, for me after that first rich experience of New Ireland, was not easily forgotten.

From time to time, therefore, during the years that followed, I was drawn to examining the few somewhat unreliable maps of the area then available in an attempt to locate the highest point of the whole island, and to muse upon possible ways of reaching it. Memories of the Danfu expedition became more cherished and as, with time, the difficulties we had experienced gradually faded, mere speculation about its feasibility became crystallised into a determination to return and complete a task left unfinished.

The only detailed map of southern New Ireland then to hand was one exhibiting a tantalisingly dense and convoluted covering of form-line contours, from which it was extremely difficult to identify the very highest ground. Furthermore, from these lines, based on an aerial survey of the forest canopy rather than representing the true ground surface, it was almost impossible to interpret the detailed relief of the very rugged interior. Only about 30 years before, at the time of the Second World War, the highest parts of the region were still being reported as lying near the western shore, with the Hans Meyer range, to the east, relatively lower in altitude. However, careful hand-colouring between the contours on a black-and-white copy of the map revealed the highest ground to be in the Hans Meyer mountains, lying just to the northeast of the Weitin and Kamdaru valleys, which diagonally divide the southern part of the island. What the map also revealed, on the complex of eastern ridges, were several patches, devoid of any lines, marked 'obscured by cloud', and lying directly on a route which otherwise offered the most promising line of ascent to the summit. My multicoloured photocopy map, which, as time went on,

As recently as the 1940s the highest part of the unexplored and rugged interior of southern New Ireland was still being reported as lying near to the western shore. However, when the dense and convoluted form-line contours were hand-coloured on this map of the region, it became clear that the highest ground was to be found in the Hans Meyer mountains just to the north of the Weitin and Kamdaru valleys. The map also indicated uncharted areas in the complex of forested ridges, marked 'obscured by cloud', through which the 1975 Kew party had to pass in order to reach the summit of the range.
Photo: M J S Sands

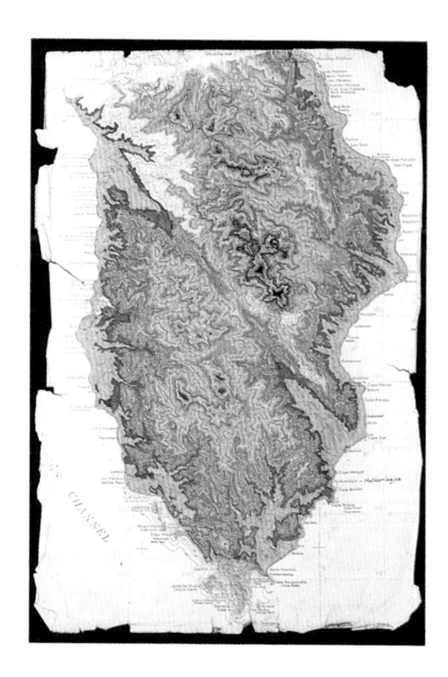

became more and more tattered like a pirates' treasure chart, was to become a constant companion during my second visit to New Ireland.

In due course, a proposal was submitted within Kew to return to Papua New Guinea with southern New Ireland as the major objective and, because plans for an expedition to Madagascar had to be abandoned funds luckily became available. Very soon I was in correspondence not only with various people within Papua New Guinea, but also, because it was then still an Australian territory, with officials in Canberra. At first, the plan to explore the highest ground in New Ireland, particularly following ridges for the most part rather than the valleys, was met with considerable scepticism. Even, much later, when the team departed from Lae to begin the New Ireland fieldwork, although enthusiasm had greatly increased, the project was still regarded by a few as too ambitious and quite likely to fail. While perhaps intended to be wise counsel, needless to say such an attitude made us all the more determined to succeed!

Arrival in Papua New Guinea

Early in August 1975 I arrived at last in Papua New Guinea after journeying for more than a week from England. With me were Jeffrey Wood from the Orchid Section of the Herbarium and Graham Pattison, then garden supervisor in the Tropical Section of the Living Collections Department at Kew. During a brief stopover in the capital, Port Moresby, we made special arrangements for through-routing consignments of living plant material by air back to Kew. We arrived in Lae the next day but for several reasons nearly a week passed before our crates of equipment were released from the customs bond-store. The delay, however, gave us an opportunity to plan in more detail the various stages of our exploration.

For Jeffrey and Graham, this was to be their first visit to the tropics and so I had thought it best to allow for periods of experience, both in rainforest and at high altitudes, before embarking on the more difficult and unpredictable New Ireland phase. Accordingly, we had intended to spend a week in the Sarawaged range to the north of Lae, but recent torrential rains ruled out the possibility and so, instead, we joined Barrie Conn of the Lae Herbarium staff on an exciting journey up the Purari River in the Gulf Province of Papua and thus our New Guinea plant collecting began.

Baimuru, a small native village in the delta swamps of the Purari, was reached by flying south from Lae through steep-sided forested valleys. The flight, in an old, single-engined 'Pilatus Porter', was made the more breathtaking by my having to hold the passenger door closed with a loop of string. That first night we stayed in a damp, wooden hotel of great tropical character, incongruously isolated among the mangroves and *Nipa* palms some distance from the village. Then next morning, with two local men, we set out in a long dugout canoe travelling for two days, first through overgrown distributary channels and then inland up the broad main stream of the Purari to Wabo, an area being considered for hydro-electric power development. We were thus able to collect quite extensively in a remote stretch of lowland rainforest before returning to Lae, just over a week later, having acquired specimens from a botanically unknown area and gained useful experience of the rainforest environment.

In the herbarium building, specimens were dried and propagating material made ready for dispatch, before we prepared to test the team in another dimension of tropical latitudes by experiencing a climb to higher altitudes. To do so it was agreed that we should drive westwards to Kainantu in the Eastern Highlands and from there journey south to Mt Piora, a mountain climbed by botanists only once before and then approached from the south with only a short stay near the summit.

Experience of collecting at tropical high altitudes was gained by climbing Mt Piora in the eastern highlands of mainland Papua New Guinea. Towards the end of August 1975, a base was established at 2000 m (6600 ft) in Habi'ina, a small village of round, thatched huts on the northern flanks of the mountain. Little of the forest below this altitude remained but just above the village, in *Nothofagus* forest, it was exciting to find a new species of ground orchid, *Calanthe cremeoviridis*, and several species of *Symbegonia*, a genus of Begoniaceae endemic to Papua New Guinea.
Photo: M J S Sands

Eventually, on the afternoon of 26 August, we set out with Jim Croft (a botanist with an interest in ferns) and local New Guineans Ozo and Beon, all of the Lae Herbarium staff, Beon driving for three hours with the clutch adroitly operated by his expansive big toe. The next day we established a base at 2000 m (6600 ft) in Habi'ina, a small village of round thatched huts on the northern flanks of Piora. Sadly, it was apparent that much of the forest below this altitude had already been removed and felling was progressing relentlessly upwards. However, just above the village, *Nothofagus* forest was still largely untouched and here it was exciting to find a ground orchid, which Jeffrey was later to describe as a new species, *Calanthe cremeoviridis*, and several species of *Symbegonia*, a genus of Begoniaceae endemic to Papua New Guinea.

After a day or two, with several hired carriers, we set out early in the morning to move up through the forest reaching 3000 m (10,000 ft) just as the light was fading. Under a rock overhang, a level platform of saplings was constructed over a mud slope, enabling a temporary camp to be made. The next day, on a nearby ridge, we collected many specimens, including several species of *Rhododendron* and, on the second morning, awoke in the cool damp mountain air to the dawn light filtering into the shelter. A hurried breakfast of porridge and biscuits was consumed before man-loads were assembled for the carriers on the next stage of the ascent. In addition each of us had our own packs.

Before long the party began climbing steeply through upper montane forest, clambering over mud, rocks and tangled roots. Some time later we suddenly emerged from the dripping, moss-covered trees onto a saddle of tussock grassland, warm in the morning sun and with distant mountains visible. Small blue gentians were to be found and in a nearby thicket there was a fine *Rhododendron culminicolum* with blood-red flowers.

Nearly two hours later in stunted forest, having cut a route round several rocky bluffs, we came out on to an open ledge where a pink-flowered *Geranium* (*G. papuanum*) was to be found amongst the grass and, over the forest canopy, there was a splendid view to the Markham valley beyond several mountain ranges. Edging along the shelf for some distance, a difficult ascent up a muddy cleft then brought us out at last on to an extensive heathland of tussock grasses interspersed with scattered thickets of the gymnosperm *Dacryocarpus compactus* and species of Ericaceae. Here, in the open, a fine array of small plants was to be found, including the pale yellow-flowered *Astelia alpina*, the fern *Gleichenia vulcanica*, and even a buttercup *Ranunculus pseudolowii*. Low-growing shrubs were also well represented, for example by *Styphelia suaveolens* and two species of *Trochocarpa*, all three in the mainly Australasian family Epacridaceae, and the rare daisy, *Piora ericoides*, previously known only from the original or 'type' collection made on Mt Piora. To the north, from this long summit ridge, a wide panorama of mountains could be seen, encompassing Mt Michael as a sharp, undulating curve, Mt Otto crouching like a sphinx and, in the far distance, Mt Wilhelm, the highest peak in the country with cloud shrouding its summit.

Eventually, after a steep and sometimes difficult ascent through stunted montane forest, the summit ridge of Mt Piora is reached and Martin Sands surveys the magnificent mountain panorama. At 3500 m (11,500 ft), tussock grasses are interspersed with scattered thickets of the gymnosperm, *Dacryocarpus compactus*, and species of Ericaceae, including rhododendrons as well as smaller plants in this subalpine environment.
Photo: J J Wood

This summit area of Mt Piora, which reaches nearly 3650 m (11,975 ft), is marked by jagged karst-like tors. They consist of a coarse conglomerate often separated by remarkable sheer-walled chasms, perhaps 250 m (*c.*800 ft) deep, their green depths filled with forest and emergent tree-ferns. On the higher open areas between the tors, isolated plants of the tree-fern *Cyathea gleichenioides* occur, frequently bearing epiphytic clumps of the orchid, *Dendrobium decockii*, which provides brilliant flashes of orange in the gloom of the swirling mist.
Photo: J J Wood

Along the ridge of Piora, at 3500 m (11,500 ft), several karst-like tors occurred which were found to be of a coarse conglomerate and on which, Eleiso, one of the carriers, took great delight in discovering an icicle! While searching for a suitable camp-site from which to explore and reach the true summit during the week that followed, we came upon some remarkable, almost sheer-walled chasms, perhaps 250 m (*c.* 800 ft) deep, their green depths filled with forest and emergent tree ferns. Nearer to hand, epiphytic clumps of the brilliant orange-flowered orchid *Dendrobium decockii* (section *Oxyglossum*) were conspicuous on isolated tree-ferns (*Cyathea gleichenioides*), which stood on one side of a shallow depression. On the other side, the depression was sheltered by a thicket and here we pitched camp. Not far away water gushed from a bank of sphagnum moss.

A week later, having reached the summit at approximately 3650 m (11,975 ft) on a very cold, wet day, the party returned to Habi'ina with a rich haul of specimens and a few days later we were back in the tropical heat of Lae.

Before leaving for New Ireland, the drying of our latest collections had to be completed and another box of living plants was dispatched to Kew. Celebrations marking Papua New Guinea's independence on 16 September further delayed our departure, but gave me an opportunity to make advance arrangements for our fieldwork on Manus Island in November. Lae Herbarium authorities agreed that Jim Croft could continue as a member of the team in New Ireland and his experience of New Guinean fieldwork was to be very much appreciated. At last, a few days later, the four of us flew to Rabaul, the old German capital at the eastern end of New Britain, which lies in a massive volcanic crater forming a huge natural harbour. Our cargo had still to arrive by sea and, as well as needing time to assemble food supplies, it was also necessary to investigate further the various possible approaches to the Hans Meyer range.

A reconnaissance flight over New Ireland at first light was essential in order to observe the mountain tops before the daily cloud build-up, and the yellow Cessna 207 in which we flew proved ideal for the purpose. Only the supporting struts of its high-mounted wings obscured a view of the ground below and there was sufficient room for all of us. For about an hour, on the selected morning, we flew at about 2200 m (7200 ft), at first through the Kamdaru–Weitin Divide and then along the eastern ridges and coastline, before climbing higher over the summit area. Here a freshwater lake was sighted briefly just before the cloud closed in. Somewhat disturbingly, the pilot chose that moment to tell us that he could not see very well because of an eye injury – but we returned safely!

The aerial observations made it clear that an approach overland from the steep and rugged west coast would be impracticable and, although a landing by helicopter on the gravel bed of the Weitin–Kamdaru watershed might be feasible, the cost of several flights would be prohibitive. Accordingly, a sea route to the eastern shore, landing somewhere to the north of the Silur mission on Cape Narum, seemed to be the best option and I set about finding a boat that would be prepared to take us. Meanwhile, Father Tony Genduza from Silur who, five years before, had been most helpful to us on the Danfu expedition, happened to be in Rabaul and, with his assistance, a radio message in pidgin was transmitted which, with luck, would alert at least some of the coastal villagers to be ready for our arrival by sea. Eventually, we were successful in negotiating for space on a small cargo vessel and, late one night with our supplies and equipment safely stowed deep in the hold, we slipped quietly away from the wharf and headed out to sea with the reflection of the harbour lights and a phosphorescent wake behind us.

Return to New Ireland

The next morning, after an extremely rough passage round Cape St George, I sat with Jim on the small forward deck as the sun came up and several dolphins arched and darted just ahead of the prow. On our port side, we could see the distant grey-blue mountains of the island rising out of a line of coastal palms and morning sea-mist, and it was at last possible to make out the shoreline and stilted houses at Natkumlagia, the village I had never really expected, in 1970, to see again. About an hour later the boat headed in shore opposite the small native village of Taron and a landing in the dinghy was attempted. However, it became obvious that the beach was too steep at that point with a fierce undertow at the surf line, so we finally anchored about half a kilometre off-shore a little way to the south.

The dinghy was lowered again and Graham was sent ashore to oversee the stores as the rest of us manhandled them out of the hold. For nearly three hours the 114 items of cargo were passed over the starboard deck-rail and despatched in a series of journeys, a standing crewman steering and powering the dinghy with a single oar lashed to

Having anchored in deep water a little way to the south of the small village of Taron, the small cargo vessel's dinghy was lowered and, for nearly three hours, stores and equipment were manhandled over the starboard deck rail and dispatched to the steeply shelving beach. Finally, members of the team swam ashore towing the floating drums of fuel linked together with a rope. *Photo: M J S Sands*

the stern. Finally the fuel drums were linked together with a rope and thrown overboard. Then, jumping into the deep water to tow them ashore, we were glad of the chance to wash off a liberal coating of peanut butter that had burst from a drum amongst the cargo.

The wooden Aid-Post at Taron became our coastal base and, after sufficient carriers had been found, an advance party began cutting a trail into the forest. Before long, with as many as 17 carriers recruited, we had established quite a substantial camp several kilometres inland. Our team of five, which included Ripano Keni, a young forest officer from Rabaul, was then able to concentrate on exploring the surrounding lowland rainforest, in which there were many very large trees including species of *Albizia, Alstonia, Dillenia, Octomeles* and *Pometia.*

In a ravine, we were delighted to discover *Christensenia aesculifolia*, a fern occurring in Indonesia but, within Papua New Guinea, only previously recorded from Bougainville. Not far from the camp, it was exciting to find quite a large lake, identified by our carriers as 'Mandih' and strongly associated with the spirits of their ancestors. A stand of betel-nut palms a few kilometres away and an immense *Octomeles* tree by the forest camp supported the view that, before the war, there were once several inland villages.

A few days later Jim and I set out from the camp at Mandih to find a route further into the mountains and, skirting the northern end of the lake, we began climbing diagonally upwards. To avoid being trapped in a valley again, we resolved, as far as possible, to follow consistently the highest ground and so, throughout that day, we cut a route steadily upwards, always keeping the Timei River to our right. In due course, through a break in the canopy, a waterfall was glimpsed issuing from the side of a high cliff ahead of us and, bearing slightly to the left, it was necessary to climb very steeply through a dense stand of bamboo.

The tall palm, *Clinostigma collegarum*, is one of several species new to science collected during the 1975 Kew expedition to New Ireland. It occurs on steep ridges in the Hans Meyer range between 1200 and 1600 m (3660 and 4875 ft) emerging conspicuously above the canopy. A group of them, left standing after clearance in front of the expedition's ridge camp at 1350 m (4430 ft), provided an impressive foreground, especially at night when silhouetted against the moonlight.
From a sketch in M J S Sand's diary

Eventually we reached the crest of the ridge and then cut a route heading ever higher to the west above the falls. The vegetation on the crest was typically more stunted and the trail rocky. In the late afternoon, at 1350 m (4430 ft), we came upon an area where though the ridge continued upwards it broadened out a little, and here, a day or so later, we made our primary and most permanent camp.

Transporting supplies and equipment up the steep inclines was often difficult, on some days being exacerbated by sickness amongst the carriers. One or two of the men, while continuing to carry heavy loads between bouts, suffered from malaria and there were several cases of serious eye infection. Frequently, Graham had to apply disinfectant and plasters to festering sores on legs and bare feet and once I had to lance an infected toe using a scalpel sterilised in the camp fire.

At first the magnificent view from the ridge camp was seen only as brief glimpses framed by foliage. However, when, as time passed, the vegetation in the immediate vicinity was cleared, the panorama gradually extended to reveal a long stretch of coastline with Cape Narum and the grass airstrip of the mission to the south and glinting flashes of the rivers in the lowland forest far below. On fine afternoons the distinct island of Ambitle in the Feni group seemed to float on an ocean haze and a number of palms left standing in front of the camp provided an impressive foreground, especially at night when silhouetted against the moonlight. They were, in fact, a characteristic feature of the ridges at that altitude, being conspicuous emergents from the canopy. Later, just before the camp was finally abandoned, one of them was felled and specimens prepared, eventually to become the type material of a new species of *Clinostigma* (*C. collegarum*), named in our honour as a team and a new generic record for New Ireland.

Collecting living plant material such a long way from any point of despatch was not an easy task, but Graham was successful in developing temporary nursery areas at both the Mandih and Ridge camps. In this way plants and cuttings could be kept in a good, fresh condition until sufficient had accumulated to make up a consignment. After a few weeks, many living specimens had been collected and Graham set off for the coast with a few of the carriers. At Taron, each item was carefully labelled and packed in a large box which was then carried for several hours along the shoreline to Cape Narum. There, it was loaded into the small mission plane and so began the next stage of its long journey to the Royal Botanic Gardens, Kew. Further consignments were despatched later.

Temporary nursery areas near the forest camps enabled plants and cuttings to be kept in good, fresh condition before being packed in a large, carefully labelled box which was carried by Graham Pattison and helpers on the first stage of its long journey to Kew. Within the box, different methods of packing are adopted depending on the type of plant, and prominent labels outside indicate that on long-haul flights the box must be stowed in a pressurised compartment.
Photo: M J S Sands

Reaching the summit of Mt Angil

It was not long after the camp on the ridge had been established that we resumed our ascent, climbing at first very steeply, and often using roots or lianes as hand- and foot-holds. Our search continued for plant specimens and for the summit areas of the Hans Meyer Range, with three more advance camps becoming operational from time to time, the most distant being a simple polythene bivouac.

By mid-October, after several weeks in New Ireland, and by stages from ridge to ridge through the forest, the team of the 1975 Kew expedition reached the summit of Mt Angil at 2340 m (7677 ft), the climb from sea-level having been achieved almost certainly for the very first time. Although much lower than many of the Papua New Guinea mainland mountains, Mt Angil, being on an island, has its highest ridges clothed in upper montane forest. Amongst stunted moss- and lichen-covered trees many plants of interest were found, including a new species of *Racemobambos*, a narrow-stemmed bamboo, and an orchid, *Dendrobium vexillarius* ssp. *uncinatum* with mauve-violet flowers, a first New Ireland record for the generic Section *Oxyglossum*.
Photo: M J S Sands

While under the forest canopy, especially as the local men had never ventured so high before, we had to rely, for direction, on the occasionally glimpsed view from a break in the cover, an altimeter, a compass – and more than a little intuition in following the convolutions of the mountains. Finally, by mid-October and by stages through the montane forest, the team reached the highest point – officially now recorded as 2340 m (7677 ft) – known to our carriers as 'Angil' and, for a week or so an advance camp was established just below the summit.

One of the penalities of following only ridges to the higher ground was the lack of any reliable source of water. At our main camp, at 1350 m (4430 ft), we constructed a long work-shelter, made mostly of bush materials but with a polythene roof which also served as an important catchment for rainwater. In addition, some water was laboriously carried up from the lower camp in an empty alcohol drum, but, when the situation became critical during a nine-day dry spell, we resorted to using water stored naturally in bamboo internodes. At our advance camp near the summit, which still took three hours to reach even after the trail had been cut, water was obtained by squeezing the dense coating of moss on the tree trunks or by sucking it up from a shallow, muddy puddle along with many of the small inhabitants. To do so, a 'straw' was used, cut from another, narrow-stemmed bamboo, which later proved to be a new species of *Racemobambos* (*R. novohibernica*) and another new generic record for New Ireland.

Although, of course, much lower than mountains like Mt Piora on mainland Papua New Guinea, those highest ridges of Mt Angil, as might perhaps have been predicted being on an island, carried upper montane vegetation which yielded many botanical rewards. With Jeffrey as a specialist, numerous orchids were collected, including, notably, a new species of *Bulbophyllum, B. hans-meyeri*, and a *Dendro-*

The discovery of *Rhododendron superbum*, with fragrant white flowers up to 12 cm (5 in) across, was a new record for this species east of the Papua New Guinea mainland. Other genera of Ericaceae found in New Ireland for the first time were represented by species of *Vaccinium* and *Dimorphanthera*. *Photo: M J S Sands*

bium with interesting mauve-violet flowers, *D. vexillarius* ssp. *uncinatum*, a first New Ireland record for the generic Section *Oxyglossum*. It was also exciting to find genera of Ericaceae for the first time in New Ireland, represented by several species but notably by *Rhododendron superbum* with white, beautifully-scented flowers up to 12 cm (5 in) across, itself the first record of this species east of the Papua New Guinea mainland.

Towards the end of October 1975, as our scheduled time of departure drew close, we began, step by step, to move the specimens and equipment down from the mountains and out towards the coast. Over the weeks we had come to know well the local men we had taken on as carriers, particularly those who had worked tirelessly to help us reach the distant camps. And so, when at last everything was packed, it was with some sadness that we boarded the government trawler *Lahara* that anchored off Taron early on Tuesday, 4 November. Not only were we leaving that marvellous, forested island but also men who we had first encountered as hired assistants but who waved farewell to us as good friends.

More fieldwork lay ahead of us in the Admiralty Islands and, for the second time, I was leaving New Ireland behind, but on this occasion, as well as another rich gathering of several thousand specimens for Kew and other institutions, a challenge had been met and the climb from sea-level to the summit of Mt Angil had been achieved – almost certainly for the very first time.

9 Orchid hunting in the Solomon Islands

PHILLIP CRIBB

On a direct route to nowhere, the Solomon Islands remain one of the least known parts of the world. They form an archipelago of seven large, 20 medium-sized and hundreds of smaller islands, lying 2000 km (1240 miles) across the Coral Sea from the coast of Queensland. To the northwest lie the large islands of the Bismarck Archipelago (New Ireland and New Britain) and New Guinea, whereas to the south the islands of Vanuatu (New Hebrides) lead down to the lozenge-shaped island of New Caledonia. The group extends 1000 km (620 miles) from Bougainville and its satellite Buka in the northwest to San Cristobal in the southeast.

Set in a warm tropical sea, these mountainous islands are for the most part covered by a mantle of tropical forest. Indeed, the generic term Melanesia is said to derive from the dark appearance afforded the islands by their continuous cover of rainforest. Many of the Solomon Islands have never been visited by botanists and few have been subjected to even superficial surveys, the rugged terrain and dense forest proving a formidable barrier to exploration to the present day. The orchid collections of earlier botanists, such as Henry Guppy, Sir Everard Im Thurn, Karl Rechinger and Stephan Kajewski, provide a fragmentary idea of the richness of the islands' flora.

Bulbophyllum dennisii, a recently described epiphytic orchid, named after Geoffrey Dennis who first collected it and who set up the botanical garden in Honiara.
Photo: Kew Photo Unit

The Solomon Islands, so long remote and virtually unknown, leapt to prominence in the Second World War when Guadalcanal witnessed one of the bloodiest battles of the entire war as the Allied forces eventually stemmed the southwards drive of the Japanese. Even so a further 20 years passed before the first organised attempt to study the geology and natural history of the islands was made. The Royal Society's expedition to the Solomon Islands in 1965 marked a new era in the botanical, zoological and geological exploration of the islands. Botanists and collectors visited most of the major islands and some of the smaller

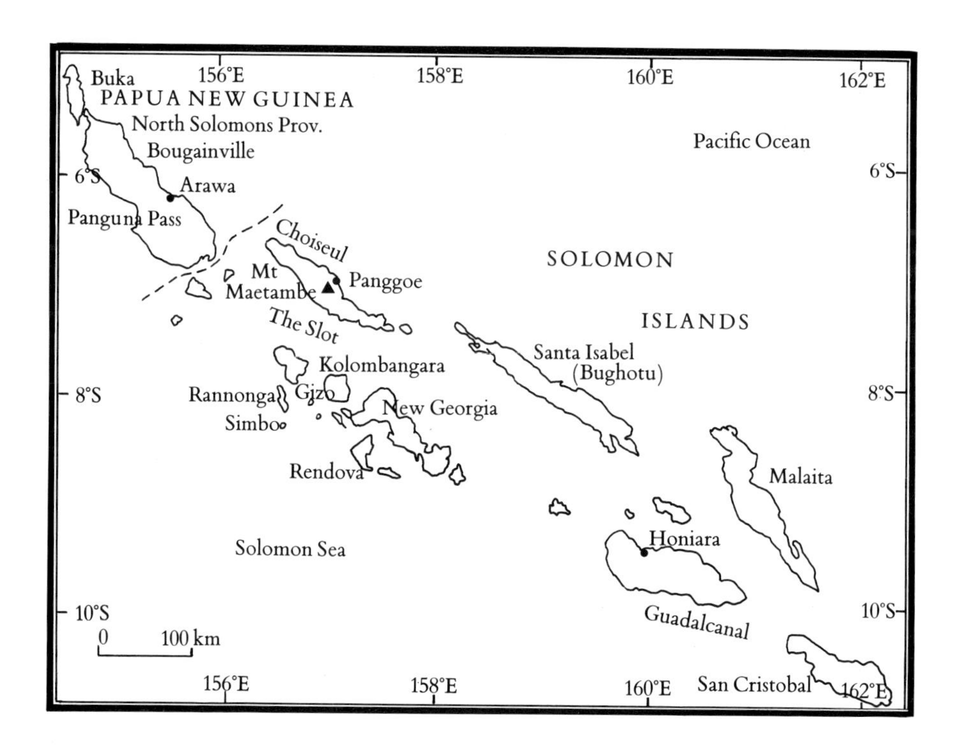

Map 11 Solomon Islands and Bougainville, with place-names mentioned in ch. 9.

ones, working from base camps and exploring the remoter mountainous interiors where Europeans had seldom ventured before.

The preliminary reports of the results of this expedition were published in 1969 in the *Proceedings* of the Royal Society. Peter Hunt, then curator of the Orchid Herbarium at Kew, participated in the expedition, collecting orchids on Guadalcanal, San Cristobal and Kolombangara. In his preliminary report he was able to estimate 230 orchid species for the Solomons and suggested that 70 of these were new to science.

Another of the participants in the Royal Society expedition was Geoffrey Dennis, an Australian, then in charge of Honiara's botanic garden. Dennis, now retired, still lives in Honiara and retains an interest and deep knowledge of the natural history of the islands and particularly their orchids. He has been for many years a regular correspondent with Kew and recently had an orchid *Bulbophyllum dennisii* named for him.

Kew's current interest in the Solomon Islands is more recent, however, stemming from a meeting in March 1978 of the International Orchid Commission in Bangkok on the occasion of the 9th World Orchid Conference. The commission passed a resolution requesting that the Royal Botanic Gardens at Kew investigate, as a matter of some urgency, the classification and naming of two groups of orchids of the genus *Dendrobium*, totalling some 90 species.

These orchids have been widely used by orchid growers in hybridising programmes to produce orchids with attractive long-lasting flowers of the type now commonly seen on sale in stores in Europe and elsewhere, imported in large quantities from Thailand, Singapore and other exotic places in the Far East. However, growers were faced with a problem when naming their new hybrids because the Royal Horticultural Society acting as orchid registrar will only accept new hybrid names if the parental names are known.

Preliminary work in the Herbarium at Kew suggested that some of the trickier problems involving the specific delimitation of variable species and possible hybridisation in the wild could only be resolved by fieldwork. By this time, I was also preparing a revision of the tropical Asiatic slipper orchid genus *Paphiopedilum*, one of the most popular groups of orchids in cultivation, and another whose known distribution extended to the Solomons. I thus had two excellent reasons for including the Solomons in my itinerary on a visit to Australasia in August–September 1980.

The Solomon Islands have been stepping stones for the movement of plants from the Malaya archipelago and New Guinea into the Pacific islands. Recent projects to catalogue the orchid flora of adjacent areas, such as Australia, New Caledonia and Fiji, have consistently shown how little we know of the orchids of the Solomon Islands. Following the success of my 1980 visit, I began to plan a comprehensive account of the orchids of the Solomon Islands. A second expedition followed in 1983, since when four further visits have been made to various parts of the archipelago by Kew colleagues and associates. In 1985, Margaret Dickson and a group of young adventurers on Operation Raleigh collected orchids for the project on New Georgia and Kolombangara; Robert and Caron Mitchell, later the same year, visited Gizo, Simbo, Kolombangara, Rendova, Rannonga and Choiseul; finally Sue Wilkinson, who is illustrating the *Orchids of the Solomon Islands*, made two visits in 1986 and 1987, the earlier visit, sponsored by the Winston Churchill Foundation and by Kew, concentrating on the Western Province and Guadalcanal, the latter on Guadalcanal and Malaita.

All of these visits have received considerable assistance from the Solomon Islands government, the Ministry of Natural Resources and Forests, the Australian Orchid Foundation, and many individuals, notably Hermon Slade of Vila and Geoffrey Dennis of Honiara. It would be impossible to provide here a complete account of all these expeditions but some highlights will serve to show the importance of fieldwork.

Into the heart of Bougainville

In the centre of Bougainville – administered by Papua New Guinea – lies one of the largest open-cast copper mines in the world. It has destroyed hundreds of hectares of forest, carving a deep pink and grey hole, bordered by lush green forest, into the mountains behind the new town of Arawa, now the capital of the province.

Arawa was a plantation owned by an Australian, Kip McKillop, until bought up by the New Guinea government shortly after independence to act as the capital of the North Solomons Province. McKillop was a keen orchid grower and gardener and the results of his enthusiasm can still be seen at Arawa. A magnificent *Amherstia nobilis*

Dendrobium macranthum, a cane dendrobium common on beach-side trees in the Santa Cruz group, Vanuatu and New Caledonia. This fine form comes from Vanikoro Island in the Santa Cruz group. *Photo: P J Cribb*

stands by the beach and was in full flower during our visit in 1980. The remaining trees in the town are festooned with orchids, notably the native antelope orchid *Dendrobium gouldii*, with showers of cream-coloured flowers, and the exotic creeping masses of *Bulbophyllum medusae*, both placed there some years before by McKillop.

The road from Arawa runs for a few kilometres along the coast north of the town before climbing inland to the mine over the Panguna Pass at about 1000 m (3400 ft). Above the fringe of coastal cultivation and coconut plantations the road runs through lush tropical rainforest until it slices through the crest of the ridge. The steep limestone banks of the pass are of raised coral full of the razor-sharp edges so characteristic of modern coral reefs. It rains every day at Panguna from midday until two or three in the afternoon, as indeed it does over most of this steamy island.

The high humidity, warm temperatures and steep open coral banks of the road have produced a habitat that is ideal for orchids. Their dust-like seeds thrive when they can germinate where there is no competition from other plants. Seeds of orchids high in the trees of the forest above the pass have blown on to the steep banks and germinated to produce a vegetation that in places consisted of almost nothing but orchids. Elsewhere club mosses, ferns and young rhododendrons were pushing the orchids aside as the forest began to re-colonise the slopes.

The aptly named *Dendrobium spectabile*, a 60 cm (2 ft) tall plant bearing spikes of three to five yellow flowers, striped with deep red, was found in small groves here. The steep slippery slopes proved difficult to climb, one attempt ending in an inelegant spreadeagled slide to the bottom of the roadside ditch and a badly lacerated arm which later turned septic. However, the struggle was worthwhile as I was able

Dendrobium mohlianum. A striking and common epiphytic orchid of the montane forest of the Solomon Islands, Vanuatu and Fiji. This specimen comes from New Georgia. *Photo: Kew Photo Unit*

Paphiopedilum bougainvilleanum, a slipper orchid which is known only from a single locality in Bougainville. It was described as new to science as late as 1971.
Photo: P J Cribb

The centre of Guadalcanal is covered in rainforest. Earthquakes are common and often sweep the forest into valley, as can be seen on the left side of this photograph.
Photo: P J Cribb

The volcanic island of Kolombangara rises to over 1850 m (5000 ft) straight out of the Coral Sea. Its upper slopes still carry fine montane rainforest but the lower slopes have been felled by recent logging operations. The orchids of this island have been studied by a number of botanists, including several from Kew.
Photo: P J Cribb

to solve an old problem of the identity of an orchid sent to me from Bougainville a year earlier by Hermon Slade, a long standing friend of McKillop. The bank was literally scattered with flowering plants of a beautiful and rare orchid, *Dendrobium rhodostictum*, with club-shaped, yellow-green stems, 30 cm long (1 ft) surmounted by two or three leaves and a mass of large pure white flowers with large flag-like petals and a lip spotted along the margin with crimson. Less common but equally widespread were the red-brown, four-angled stems of its close relative, *D. ruginosum*. This has less flamboyant, smaller white flowers, with narrow petals and a narrower lip veined with dull green. Suddenly, I found a plant in flower that reminded me of Hermon Slade's orchid. A close examination showed that it was intermediate in all its characteristics between *D. rhodostictum* and *D. ruginosum*, in fact, a hybrid and not a distinct species.

That afternoon we visited the mine, whose true extent was painfully visible from the top of the pass. The visit provided an unexpected and welcome bonus, however, for the manager very kindly offered us the use of his helicopter at dawn the next day so that we could get high into the mountain forests and then have a whole day to walk out, exploring as we went. We gladly accepted and the following morning were up well before dawn, taking off from the mine just as light filtered onto the mountains over Panguna. A first helicopter flight is, I suppose, always a memorable experience but being flown within yards of high cliffs, shaggily green even on near-vertical exposures, was quite exhilarating!

One of the main aims of our expedition was to see one of Bougainville's rarest plants, the lady's slipper orchid, *Paphiopedilum bougainvilleanum*, in its native habitat. The helicopter flight suggested suitable localities and armed with permission from the chief minister and local village elders we headed for a likely site high in the mountains. Walking or scrambling uphill has never been my forte. When the paths are greasy, the vegetation soaked and the slopes approach 45 degrees it can become a nightmare of endurance. However, after four hours we reached the base of a series of massive, rocky outcrops in the forest. The first yielded little but a treacherously narrow skirting track, around which we edged with some nervousness, clinging to saplings rising from below on the near-vertical slopes of the ridge. The second outcrop, looking from below like limestone, rose 100 m (330 ft) out of the forest and effectively blocked any further progress upwards.

Closer inspection showed that the whiteness of the rock was caused by lichens growing on the wet surface of a boulder of volcanic origin. Around its base and among sedges and moss, a careful search revealed a few plants of the slipper orchid. We were only the second Europeans to see this delicately coloured orchid in the wild. Kip McKillop and Hermon Slade had discovered it in the late 1960s and it had been described as recently as 1971. Sadly, recent depredations by plant collectors have almost exterminated it from all of its known localities on the island.

New orchids from Choiseul

When Henry Guppy visited the large island of Choiseul (pronounced 'Choysul') in September 1883, he found that the inhabitants of Choiseul Bay gave him a wide berth, scarcely surprising as two years before HMS *Emerald* had exacted reprisals on the village of Kangopassa, whose inhabitants had earlier massacred some of the crew of the trading vessel *Zephyr*. Guppy made an excursion some way up one of the rivers which flow into the bay, accompanied by a local chief, Krepas, and his son Kiliusi. Guppy remarked in passing in his subsequent book that Krepas was a practising cannibal!

Until 1986, Guppy alone had collected plants on Choiseul and so the visit of Robert and Caron Mitchell to the island could be truly described as a pioneering one. Robert, a horticultural student at Kew from 1983–6, had won the Ernest Thornton-Smith Travel Scholarship awarded each year to the Kew student who put forward the best idea for a visit abroad to study plants. During the early work on the Kew project to produce an orchid flora of the Solomon Islands it became obvious that the orchids of Choiseul, as indeed the rest of its flora, were completely unknown. Robert planned his expedition to remedy that situation.

Throughout the Solomon Islands and, for that matter, in most other parts of the world, land belongs to people. In the Solomons, the land and the trees are all owned by villagers. Permission had to be obtained from the government and the local people to enter the forest and to collect plants there.

The Mitchells based themselves at the village of Panggoe, halfway down the northeast coast of the island. From there, they aimed to climb Mt Maetambe (1006 m, 3300 ft), the highest peak on the island. The assault on Mt Maetambe began at the coastal village of Ghaghara, just south of Panggoe. The narrow coastal areas have been largely cleared of forest for coconut plantations but inland the terrain is very dissected by high ridges which run off the central mountain chain to the coast, following the courses of the rivers and streams. The ridges are covered by rainforest little affected by man, but the valleys contain dense stands of *Heliconia*, a beautiful wild long-stemmed banana, palms and gingers.

The heavy rain that descends on these mountains each day can reduce visibility to a few metres, yet a few kilometres away the coast can be bathed in brilliant sunshine. Despite the outrageous conditions on the mountain, the Mitchells collected 45 species of orchid on the ascent. All proved to be new records for the islands, the most interesting being the terrestrial orchids, the primitive *Tropidia disticha* and the pale ghostly leafless *Stereosandra javanica*. The latter lives on decaying organic matter (a saprophyte) and was particularly exciting, being a new generic record for the Solomon Islands.

The return journey down the mountain proved difficult as the gentle streams encountered on the ascent had turned to raging torrents and strong winds had brought branches crashing down to block the narrow track. As Robert Mitchell wrote in his report he was 'left with a new respect for the island and the people that live there'.

Lord Howe Island between Australia and New Zealand. This view of the mountains is taken from Transit Hill in 1985 during a plant-collecting visit while working on the *Flora* of the Island.
Photo: P S Green

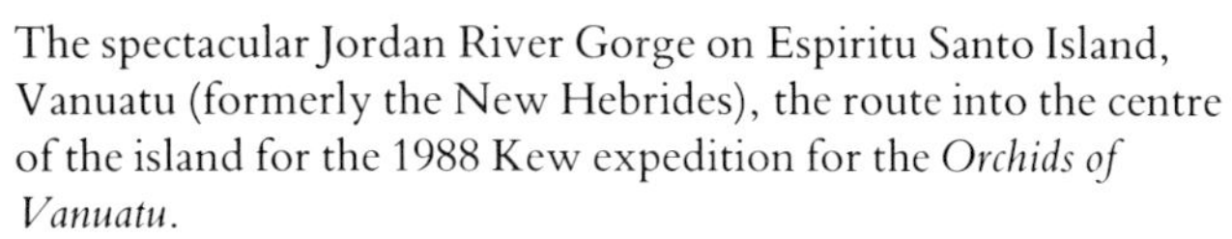

The spectacular Jordan River Gorge on Espiritu Santo Island, Vanuatu (formerly the New Hebrides), the route into the centre of the island for the 1988 Kew expedition for the *Orchids of Vanuatu*.
Photo: P J Cribb

The expedition guides on Espiritu Santo, Vanuatu, wore headdresses of *Selaginella* gathered at the top of Mt Tabwemasana.
Photo: P J Cribb

The Napier Range near Windjana Gorge, Western Australia, part of the remote area studied for four months in 1988 by a multi-disciplinary British-Australian Kimberley expedition. Mounted jointly with the Royal Geographical Society, the project marked the bicentenary of the Linnean Society of London and was also a major event in Australia's bicentennial year. Martin Sands was the Kew botanist and deputy leader.
Photo: M J S Sands

This attractive species *Begonia merrittii* was collected by Martin Sands during his visit to the Philippines in 1976. During his field study of the ecology and distribution of Philippine begonias, the presence or absence of some species proved useful in assessing the condition of remaining primary forest and in giving an indication of its value for conservation. *Photo: M J S Sands*

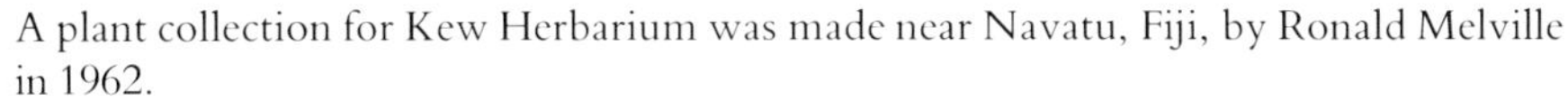

A plant collection for Kew Herbarium was made near Navatu, Fiji, by Ronald Melville in 1962.
Photo: R Melville

During Ronald Melville's 1961–62 stay in New Zealand collecting plants for Kew and the Christchurch herbaria, he visited Lake Wakatipu near Queenstown. The mountains known as the Remarkables lie beyond; although large areas are planted with northern-hemisphere conifers, native *Hebe* and epacrids survive at higher levels.
Photo: R Melville

Lady's slipper orchid on Guadalcanal

In camp on Guadalcanal John Campbell, a New Zealander who joined the Kew orchid expedition, drinks morning tea outside the sleeping-hut in Sutakiki Valley.
Photo: P J Cribbs

In the deep interior of Guadalcanal is an orchid with a price on its head, a rare lady's slipper orchid, a species desired by the connoisseur orchid-grower just as fervently as the keenest collectors of porcelain or works of art. Geoffrey Dennis, then in charge of the Honiara Botanic Garden, sent a dried herbarium specimen of this slipper orchid and a photograph of it clutched by a Solomon Islander to the herbarium at Lae in New Guinea some 20 years ago. Geoff had chanced across this plant while botanising in the mountainous interior of Guadalcanal. Aware of the significance of his discovery, he returned several times to that area over the next few years but the Guadalcanal slipper orchid had disappeared and the area where he found it had been converted into a maize garden.

In 1980, Alasdair Morrison and I joined Geoff Dennis in a futile search for the Guadalcanal slipper orchid. However, in 1983, the exciting news reached Kew that Geoff had rediscovered it away from the vicinity of his original collection. The timing of his news was fortuitous because I was, at the time, preparing a definitive account of the genus *Paphiopedilum* and was keen to include the Guadalcanal species, one of the few species I had not yet seen in the living state.

I flew into Honiara in August 1983 to be met by Geoff Dennis and another good friend and orchid enthusiast, John Campbell, a tough New Zealander familiar with the mountains and forests of the Solomons. One week later we began our search for the orchid with the necessary government and local permissions. The first day's march into the interior of the island was gruelling, carrying a backpack up steep slippery slopes through dense forest and over two mountain summits. Fortunately, the abundance of orchids in these forests gave me plenty of excuses to stop and botanise en route, vital for aching limbs!

At the top of the first mountain, the tall forest gave way to a dwarfer forest draped in dense moss. Buried in the mossy trunks and branches grew a variety of orchids: the jewel-like miniature *Dendrobium minutum* with tiny cornet-shaped scarlet flowers, the closely related but slightly larger pale purple and red trumpets of *D. laevifolium* and twining with them up the trunks in a wide spiral a large *Bulbophyllum*. A short search produced two flowering spikes of the latter, each with a club-shaped head of large purplish flowers. This, on subsequent examination, has proved to be new to science.

On the descent into the valley separating us from the main mountain, I passed under a branch at head height above the slippery track and, watching my feet, almost missed the hanging necklaces of snow-white flowers of *Coelogyne veitchii*, perhaps the prettiest of all Solomon Islands orchids. This had previously only been recorded from New Guinea.

Our base camp was established in a vacant branch-walled and palm-thatched hut in a small village in a valley deep in the forest. Running water was provided by the local raging river and a shower could be taken under a low waterfall – the height of luxury after a long exhausting day in the forest.

An early rise before dawn the following day took us into the forest again up slopes even more precipitous than the previous day. In the absence of a path we clambered up using the branches and trunks of trees for support, eventually emerging on a narrow ridge scarcely wide enough in places to walk along. The frequent earth tremors and earthquakes had caused numerous landslips in the area and the slopes had been partially cleared of forest. Eventually the ridge broadened out and we entered a taller forest where, to our delight, we found a small colony of the slipper orchid. A diligent search of 200 plants revealed half a dozen or so in full flower. Above dark and pale green mottled leaves and atop a slender tall stalk was borne a 10 cm (4 in) flower with deep glossy purple petals and glossy green-brown porch. Indisputably, this orchid was *P. wentworthianium*, known from just two places on earth. It had been discovered on Bougainville in 1961 by Kip McKillop, who gave plants to a visiting American orchid grower, Clayton B Wentworth, after whom it was named in 1968. A few plants of this variety were removed with the permission of the government and the local landowner. These have subsequently flowered at Kew and the seed produced has recently germinated well. Hopefully, these plants will form a stock from which *P. wentworthianium* can be introduced into cultivation without the wild population suffering the depredations of collectors.

Ascoglossum calopterum, an attractive epiphytic orchid still found on trees within the town boundary of Honiara.
Photo: Kew Photo Unit

Further prospects

What progress has been made in Kew's work on the orchids of the Solomon Islands? So far the accounts of both of the sections of *Dendrobium*, requested by the International Orchid Commission, have been published, in 1983 and 1986 respectively. The revision of the genus *Paphiopedilum*, including the Solomon Islands' species, was published in 1987. A preliminary checklist of the orchids of the Solomon Islands was produced in 1983 and the forthcoming book *Orchids of the Solomon Islands* will be of interest to botanists, horticulturalists and conservationists. Even in the few years since I first visited the islands, logging has destroyed a considerable proportion of the forests on many of the islands. Reserves and even national parks are urgently needed to preserve the diversity of the region's vegetation.

A knowledge of the orchid flora can provide a sound basis for discerning the species-rich areas most needing protection. Orchids are susceptible to forest destruction and are an ideal gauge of environmental disturbance. We hope that the orchid book will highlight the richness of the remaining Solomon Islands' forests and provide the stimulus for the protection of at least the most interesting of them.

Paphiopedilum wentworthianium. This rare slipper orchid was recently discovered in the interior of Guadalcanal by Geoffrey Dennis who is a regular correspondent with Kew. Its only other known locality is on Bougainville.
Photo: P J Cribb

10 *From coast to campo in Bahia, Brazil*

SIMON J MAYO

In recent years, Kew has built up a strong research programme in Brazil. This has resulted from our good fortune in establishing bilateral collaboration with various institutes there, first with the Cocoa Research Centre and later with the University of São Paulo and the Botanic Garden of Rio de Janeiro. The first projects were to carry out basic survey work in areas which had been little explored by botanists, but later the aims were refined and expanded. Today the objectives of Kew's joint research programme with Brazil include the preparation of a detailed inventory of the plants of the *campo rupestre*, a montane vegetation-type found on outcropping rock mainly in the interior of the states of Minas Gerais, Bahia and Goiás, and the study of the flora of the state of Bahia as a whole. Through a formal link with the University of São Paulo, other lines of work are now being developed.

Like most Kew botanists involved in the classification of South American plants, my formative experience of Brazil was gained during an expedition to Bahia and the following account is based on that trip, which took place between January and April 1977. Before embarking on this, however, some background is necessary.

Kew's involvement in field research in Brazil has a long history. The most important botanical study of the 19th century in Brazil was carried out by Richard Spruce, who spent 15 years on the Amazon and in the Andes between 1849 and 1864, and made a collection of outstanding importance. Kew's most famous Brazilian connection is probably the episode of the rubber seeds, collected in Amazonia in the 1870s by Henry Wickham at the instigation of Sir Joseph Hooker, then director of Kew. After languishing in Asia ignored for several decades, plants grown from Wickham's seeds eventually formed the basis of the Malaysian plantations developed and controlled by the British imperial administration. Free of the limitations imposed by endemic Amazonian

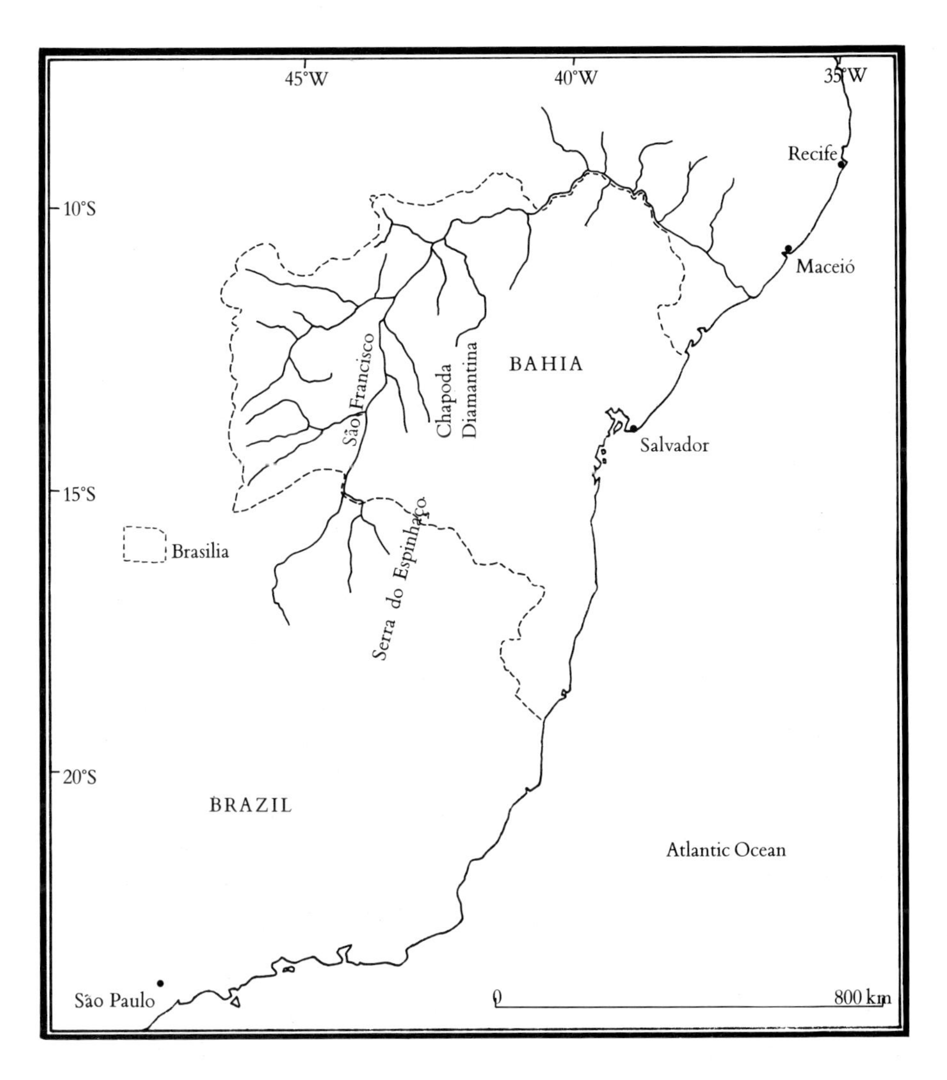

Map 12 Eastern Brazil with Bahia State, and places mentioned in ch. 10.

pests, these plantations were extremely productive and eventually led to the demise of the great Amazonian rubber boom.

From the turn of the century until the 1960s, Kew played no major part in the botanical exploration of tropical America, concentrating instead on British colonial territories in the Indian subcontinent and tropical Africa. However, a change in focus began after a two-year UK–Brazilian expedition to northern Mato Grosso organised by the Royal Society and Royal Geographical Society of Great Britain and the Brazilian Academy of Sciences. This major multidisciplinary project was concerned with the scientific exploration of a region which was only then being opened up by road-building. Its impact on British scientists was significant and many long-term joint research projects developed subsequently as a result of this experience of the country's extraordinarily rich natural environment.

One of the British botanists involved was Raymond Harley, then only recently appointed as head of Kew's South American section, and during a later expedition in 1971 to Central Brazil, he made his first

Raymond Harley with *Paepalanthus speciosus*, his 20,000th collection, made near Rio de Contas.
Photo: S Mayo

acquaintance with the flora of the northeastern state of Bahia. At that time the flora of this region, particularly in the interior, was very poorly known. Even the coastal region was not well explored botanically, despite the fact that the state capital, Salvador, is one of Brazil's oldest ports and invariably a stopping place for Europe's travelling naturalists of the 19th century, including Charles Darwin. The coastal region in the south of the state, where the richest natural forest occurs, was difficult to penetrate until the 1970s, when the first arterial coastal road was built: the trade of coastal towns like Ilhéus with Salvador was mainly by sea until the 1950s.

By the late 1960s, the Cocoa Research Centre (CEPEC) at Itabuna in southern Bahia (CEPEC is the research institute of CEPLAC, Brazil's federal agency for cocoa agriculture) had begun more intensive botanical exploration of the coastal region. With skilled field personnel, a good herbarium and excellent support facilities, CEPEC was an ideal base and through the interest and enthusiasm of Paulo Alvim, then CEPEC's director, an international collaboration was established with Kew. So it was that in 1974, Raymond Harley led the first Kew–CEPEC expedition to Bahia. This was a great success and the results showed how extraordinarily interesting was the flora of Bahia. Many species were discovered that turned out to be new, and even some new genera. International specialists who had been asked to identify specimens of their chosen plant groups were by turns perplexed and wildly enthusiastic. With such exciting discoveries in such abundance it was obvious that further fieldwork should be carried out, and the rapid disappearance of the coastal forest lent urgency to the task.

It is an unfortunate fact that most of the southern Bahian rainforest, packed with rare and endemic plants and animals, disappeared in a few short years in the 1970s, immediately following the completion of the coastal road linking Rio de Janeiro and Salvador. This happened at the very moment that biologists were beginning to understand how important the Brazilian coastal forests were as a genetic reservoir. The same sad story can be told of the forests from Pernambuco in the north to Santa Catarina in the south, and it may be said without exaggeration that the destruction of the coastal rainforests of Brazil ranks as one of the great biological disasters of modern times.

The coastal forest

At Raymond Harley's suggestion I first went to Brazil in the last few days of 1976, joining him and Ruth Henderson, then supervisor of the Palm House, as part of the Kew contingent. This second Kew–CEPEC expedition, also led by Ray, was aimed particularly at the flora of the coastal forest (Mata Atlântica) and of the mountain ranges of the interior, where *campo rupestre* vegetation occurs.

Rio is a good place to start a first visit to a tropical country. The breathtaking scenery of mountains, coastline and forests surrounding the city makes a strong and lasting impression, calling to mind Darwin's lines on first experiencing tropical forest at Salvador in Bahia:

'The day has passed delightfully. Delight itself, however, is a weak term to express the feelings of a naturalist who, for the first time, has wandered by himself in a Brazilian forest . . . '. The city lies around the southwestern margins of the huge lagoon-like Guanabara Bay, and spreads west along the coast, now rocky with striking steep-sided hills like the Sugar Loaf, now opening out into great white strands – the famous beaches of Ipanema, Copacabana, Leblon, and Flamengo. The hills tower above the city, and to the west are covered with forest. The city boasts its own rainforest national park, the Floresta de Tijuca, which in places reaches right into the inner suburbs.

We were very fortunate to stay for the first two nights with Margaret Mee, the English botanical artist (1909–88), and her husband Greville, who were old friends of Ray from previous trips, and who welcomed us warmly at their house, perched on the edge of a steep wooded slope in the beautiful old district of Santa Teresa. There, with a cup of tea and a boiled egg, I soaked up my first impressions of the tropics; at nine in the morning the heat was already great, the air filled with the tremendous penetrating sound of cicadas, which is exactly like a band saw, constantly swelling and fading away. There was the intense green of weedy elephant grass growing up the roadbanks, gigantic jackfruits hanging overhead, and hummingbirds, alarming at first as they whizzed past one's head, sounding like enormous hornets. After grappling for a while with these unfamiliar sensations I began to pick out the forms of aroids, bromeliads and orchids; the garden was bedecked with plants, many being collections made by Margaret during her Amazon expeditions and used for her wonderful paintings.

After two days of visiting botanists and botanical institutes in Rio, we caught a bus to Bahia, which took us along the BR101, the arterial coast-road which had given rise to so much forest clearance. By early on Sunday 2 January we were in Bahia, peering sleepily out of the bus windows at a somewhat depressing vista of burnt, smouldering forest remnants. The larger trees still standing were, however, covered with epiphytes which promised well.

The following nine days were spent making various necessary arrangements for our fieldwork. The first problem to be sorted out was the release of our equipment from the customs in Salvador, where it had been sent from London by sea several months previously. In resolving this problem, and many others, the help of Paulo Alvim and his staff at CEPEC was invaluable. In between the various chores we also found time gradually to settle in, getting to know our new colleagues and seeing a little of the surrounding country and nearby towns.

The CEPEC institute at Itabuna was a modern building set in very large grounds divided into numerous plots, each supporting different varieties of cocoa, under a variety of trial regimes. The main aim of the scientific work there is to improve the cultivation of cocoa by all means available, which includes breeding and other types of agronomic improvement. However, for the most part, cocoa is still grown as an understorey tree-layer below large standard canopy trees, the remnants of the original forest. Thus a smattering of the original tree flora remains in cocoa plantations, as well as many of the epiphytic plants. Recent research has linked the biology of the pollinating insects of cocoa to epiphytes and the presence of the latter may be essential for successful set of the cocoa pods.

Itabuna was the nearest and largest town of the district. In 1977 it was undergoing rapid development and like many growing Brazilian towns, wore a rather untidy air, a smelly bustling sort of place with every kind of shop and workshop imaginable. At the centre brash new office blocks elbowed out the older and humbler one-storey buildings, testimony to the cocoa trade of the region, at that time extremely successful. The market impressed us most, constructed like a kind of amphitheatre, with the fruit and vegetables occupying the centre and encircled by rows of stalls selling meat and fish. Huge piles of large green oranges lay on the ground and the stalls were loaded with all manner of tropical fruits, fascinating to our novice eyes.

Ilhéus by contrast was the old, historical town of the region, lying at the mouth of the Rio Cachoeira, where it meanders into the Atlantic, fringed with mangrove. This town was made famous by the novels of Jorge Amado, who wrote in particular of the life of the cocoa region in its pre-war days, when establishing a plantation meant cutting it out of the primary forest by hand in then remote areas. Tough landowners, known as 'colonels', established dynastic family-holdings, and were accustomed to resolving their frequent territorial disputes by violent means.

Ilhéus is also well known to botanists who study neotropical plants as it was a classic collecting locality for many European naturalists, who visited the area by taking a boat south from Salvador. Of particular interest to me was the journey of Archduke Ferdinand Maximilian of Austria. He made a very successful expedition in the

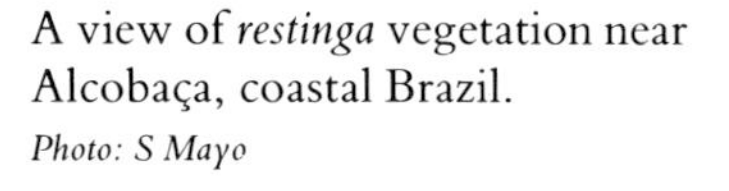

A view of *restinga* vegetation near Alcobaça, coastal Brazil.
Photo: S Mayo

mid-19th century and near Ilhéus collected numerous new species of the arum lily or aroid family (Aracae), my own special interest. Many were later superbly illustrated in Vienna in a magnificent folio entitled *Aroideae Maximilanae* ('Maximilian's Aroids'!).

After the first day or so we ensconced ourselves in some simple rooms at the institute and learned how to cook local food. We discovered that the climate necessitated new habits. One particularly fine stew cooked by Ruth was left for the following day's dinner, but on opening the pot later we discovered it was literally bubbling with putrescence and had to be consigned to the cocoa plantations while we covered our faces to avoid the noisome stench. At night we were regaled with the most extraordinary orchestra of frogs and toads. Walking out into the surrounding plantations after dark we were assailed by a pulsating 'wall' of extraordinary sounds, full of intricate details and shifting rhythms with all voices present, from soprano to contrabass, and timbres which varied from forthright clonks to eerie, almost human squeals.

We set out at last on our first collecting trip on Ray's birthday, 11 January, a band of seven people in two vehicles. Our CEPEC *companheiros* were the plant collectors Raimundo Pinheiro and Talmon dos Santos, and drivers Gilberto and Augusto. Since the rapid growth of the CEPEC Herbarium during the late 1970s and 1980s, Talmon dos Santos has become a well-known plant collector. Both he and Raimundo were tough, hardworking and very knowledgeable, well-versed in the craft of making good botanical specimens in forest. But they were quite different characters.

A striking member of the pineapple family *Billbergia porteana*. This family, Bromeliaceae, is almost completely restricted to the New World.
Photo: R M Harley

Talmon was built like a weightlifter, and immensely strong; he liked to show off his strength and once impressed us by lifting an old British Navy cannon from its carriage on the beach at Porto Seguro! But he was a gentle giant, a kindly soul who would stride into the square of the poverty-stricken villages we passed through in the interior and hand out money and cigarettes spontaneously. Raimundo Pinheiro was older, a natural leader who after a few beers would talk with gusto of the wild days of his youth, when he seemed to have been a local terror. He took pride in his botanical field knowledge, and could identify most trees with their common names and botanical families. He had accompanied Ray on the 1974 trip and, though not employed by CEPEC when we arrived in 1977, discovered somehow that Ray was back and turned up at Itabuna to offer his services again.

Our first objective was the national park at Monte Pascoal, created initially as a national monument. The hill which gives the park its name is said to be the first sighting that the Portuguese explorers had of Brazil when they 'officially' discovered it in 1500; it being Easter Day (Páscoa), the hill was named 'Easter Mountain'. Although recently reduced in area to accommodate the needs of the remaining indigenous Pataxós Indians, the forest of Monte Pascoal is still the largest remaining fragment of the great rainforests of southern Bahia. It is thus of outstanding importance botanically and much still remains to be done there. Our two-day visit was but a very brief taste of its riches, but sufficient to see some extremely large trees and many interesting herbaceous plants, especially aroids. One of the most useful collections made here was of the palm *Polyandrococos caudescens*, which back at Kew turned out to have been previously collected very few times.

After spending two nights in the simple house which served as a field station in the park, we moved some way south to the coastal town of Alcobaça to look at the vegetation known as *restinga*. The word is loosely used for a complex of vegetation types, ranging from high forest to the merest straggle of ground herbs. A constant feature, however, is the sandy soil.

Restinga is found throughout the eastern seaboard of Brazil and is perhaps the most threatened of all Brazilian vegetation-types because it occupies land which, with the development of coastal towns and cities, commands the highest prices – everybody wants a house on the beach! In 1977 the *restingas* of southern Bahia were being rapidly replaced with housing developments, but CEPEC vegetation scientists had shown us maps of the best remaining areas for botanising, and the vicinity of Alcobaça boasted one of the largest. Work in Bahia has shown that though many species of *restinga* are found in some of the inland forest types, there are all sorts of plants which are either endemic to *restingas* or else show very odd disjunct distributions – for example, they may occur also in the mountains of the interior, but nowhere in between.

This kind of information is very important, not only for building up a picture of present-day plant distributions, but also for attempting to discover how the vegetation has changed in recent geological time. Work by Brazilian geographers has suggested that during and since the Ice Ages of the northern hemisphere, large areas of Brazil which were covered in dense forests in modern times (i.e. before man got to work with axe, matches and chainsaw) were periodically covered instead by more arid types of vegetation. When the climate became moister again, some of the plants adapted to these open environments may have become 'trapped' in the coastal *restingas* by the growth of an intervening zone of tall forest.

At Alcobaça we made our first real camp. Our gear consisted of four two-man tents, in pale boy-scout green, and a large red tarpaulin affair which Ray had had specially made for the previous Bahian expedition. The idea was to provide a shaded and rainproof working area in which the drying frames could be installed, work on the plants carried out, the cooking done and meals eaten. Although requiring a clear head to remember how all the various poles fitted together, this system worked very well throughout the trip. Ruth and I, the novices, began to learn the working routine in the field. Rising generally at 6 a.m., the first job after breakfast (coffee and porridge made by Raimundo) was to examine the pressed plants and remove those dried overnight.

The plants were pressed between sheets of thick absorbent paper and aluminium corrugates and kept tight by press frames fastened by ropes. The whole package, or rather several such, would be laid sideways on top of the drying frame. This is a simple contraption –

essentially a large wooden box lacking top and bottom with, in our case, kerosene stoves on the ground inside. Once assembled, the stoves create an upward draught of hot air which passes through the corrugates and thus dries the plant specimens. When working well the system can dry plants in a few hours and the more rapid the drying, generally the better-looking are the specimens.

A few essential points of technique are needed, however, to get the best results. The most important of these is that the presses have to be regularly tightened as the plants dry otherwise the specimens start to crinkle, which not only looks bad, but also makes them more prone to damage later on. Since there were five of us collecting every day and we were trying to collect every plant in sets of 10 duplicate specimens for donation to various institutes, we needed a lot of equipment; four drying frames and enough drying paper, corrugates and other tackle to keep them running 24 hours a day.

After checking the presses, we would sally forth at around 8 or 9 a.m. in one of the vehicles to find a likely spot to investigate. Once at a good locality we would all go our separate ways into the *mato* – a useful word to describe any kind of vegetation – with machetes, plastic sacks and cameras, eventually congregating to examine our prizes. A useful sorting out of material generally took place at this point, and we would find that several people had collected the same thing, or that we needed more specimens of a particular plant. Ray would then start to make the field notes, while we would grab a plant to press and ask him for a number. The field number, the most essential reference for the specimens, would be scrawled on the flimsy papers in which we pressed the individual specimens and then the plants would be put into a press. After a simple lunch of biscuits, guava jam and perhaps a bit of cheese, we would continue work until 4 or 5 p.m., by which time we usually had a lot of material to deal with.

We generally returned to camp via some roadside bar, where we would take a welcome beer or iced coconut milk after a day in the sweltering heat, and so to the evening's work of preparation. Once again the presses on the drying frames had to be examined before putting the day's collections between corrugates and drying paper. Further notes would be made, plants preserved in alcohol and living collections written up, and most important, the newly collected plants would be rearranged in their flimsies to give the best disposition when dried. We broke for dinner between 7 and 8 p.m. by which time we were generally famished, and the work would proceed late into the night. The last job, which Ray and I shared, was to write up the diary of the day's doings, tighten the presses for the last time and sink thankfully into bed at midnight or 1 a.m.

The *restinga* was full of botanical interest. Its most typical form consisted of scattered small patches of low forest often with many pineapple relatives (bromeliads) and aroids growing beneath the trees, separated by open areas in which trailing plants and even cacti grew. Orchids were common; in one site Ruth found three species of *Catasetum*. In wetter areas marshy vegetation occurred, rich in sedges

Philodendron williamsii, a magnificent member of the arum family (Araceae) in the Bahian *restinga*. This family is very well represented in South America and is the subject of studies both in the field and at Kew Herbarium.
Photo: S Mayo

(Cyperaceae), grasses (Gramineae), xyrids (Xyridaceae) and pipeworts (Eriocaulaceae).

Four days later we moved north to look at the forest again at various localities. The most interesting was near the town of Una, where we found lodging in the midst of a rubber plantation. One night there was memorable for a tremendous thunderstorm, through which Raimundo slept very soundly with his radio playing at full blast next to his ear. I remember switching it off at about 3 a.m. The forest was exciting and full of unusual plants. Later field research carried out by CEPEC and collaborators has confirmed that this region is a 'hot spot' of endemism, not only in plants but also in many animal groups. It harbours an important primate reserve, one of the last places where the almost extinct Golden Lion Tamarin occurs. Among many other plants, two species that stand out in my memory were the second collections ever made of a beautiful little *Philodendron, P. recurvifolium*, previously found only by Ferdinand Maximilian a century before and *Becquerelia clarkei*, a handsome forest sedge with broad leaves tinged with deep purple on the underside.

Our next stop was the seaside town of Itacaré, where Ray was anxious to collect more material of a tree which he had collected there in 1974 and which seemed to be an extraordinary new genus of the Leguminosae. The town lay at the mouth of the Rio de Contas and was reached by a rough road that wound through hilly country, much still covered with forest and cocoa plantations. Our sojourn there was extremely pleasant; conditions seemed ideal for the tropical botanist. Our camp site was a grassy knoll a stone's throw from a beautiful beach, where great Atlantic rollers beat their constant rhythm. A tiny shack of a bar (*boteque*) stood right opposite and was always open when the need was greatest; the local town council had thoughtfully provided freshwater showers on the beach to wash off the salt resulting from sea bathing; a stream meandered down from the forest through a coconut grove to the sea, so that our slumbers were accompanied by a chorus of amorous frogs. Last but not least, the surrounding forest was packed with amazing plants.

Harleyodendron, the new legume genus, was soon rediscovered in full flower and occasioned much rejoicing. The discovery of a new species, while always gratifying, is not such a great novelty for botanists working in the neotropics, where so much exploration still remains to be done, but new genera are more unusual. This one looked nothing like any of the plants to which it is now known to be related, and as we examined it I remembered how the original specimen collected in 1974 was handed from one eminent specialist to another in the tea-room of the Kew Herbarium, accompanied by much wagging of heads and pursing of lips – 'Well, I suppose it might be a legume' was the sort of comment one heard.

Our 25-day coastal collecting trip was rounded off by a visit to another *restinga* site north of Itacaré, near the little town of Maraú. The high point for me was a huge and spectacular aroid, *Philodendron williamsii*, growing in splendid isolation in the open *restinga*. Further research in herbaria from all over the world has yielded virtually no other field collections of this species, although this is perhaps not so surprising in view of the difficulties of reducing such a gigantic herb to a neat, manageable herbarium specimen. The glossy club-like fruit-stalks (infructescences), filled with a brown gelatinous secretion, took over a month to dry!

I returned to Itabuna feeling already like a veteran, though in fact I had only barely begun. Two things I had learned were that I had a hitherto unrecognised ability to eat huge quantitites of roast meat, something that was already becoming an expedition joke, and second that if we were to collect the really awkward plants, like aroids, palms and giant saw-edged Cyperaceae, then I had to do it, because they did not seem to appeal to anyone else.

Kielmeyera species, a beautiful shrub in the family Guttiferae from the *campo rupestre* of Diamantina, Minas Gerais state.
Photo: S Mayo

Into the interior *campo rupestre*

After about a week organising our collections, we set off for the interior to collect in the *campo rupestre* vegetation of the mountains. Central Bahia consists of a large rugged massif, reaching altitudes of 2000 metres (6500 ft) in places, which because of its diamond-bearing ores is known as the Chapada Diamantina. Gold and precious stones were discovered in the 18th century, and resulted in the establishment of a series of pretty little towns along the eastern edge of the massif. The heyday of diamond prospecting in this region is long since past, and today these towns rely on tourism and the collection and trade of everlasting plants from the *campo rupestre* vegetation that surrounds them. In 1977, Andaraí and Mucugê, which we visited first, seemed to be far too big for the number of people living there, and the atmosphere was of a sleepy, mouldering decadence, with grass sprouting between the cobbles of the squares and the elegance of the houses suggesting some earlier golden epoch.

The term *campo rupestre*, like *restinga*, is generally used rather loosely. Other vegetation types are typically found in association with it, such as upland marshes and bogs, grassy downs, forests along water courses and flat areas with various types of eroded stony substrate. The tree lilies (Velloziaceae) are very characteristic of *campo rupestre*, and two other plant families which are particularly rich are Eriocaulaceae and Xyridaceae, the most important components of the everlasting-plant trade because of their suitability for dried flower arrangements and the delicate beauty of their inflorescences. The species are extremely numerous and often confined, or so we believe, to very small areas. Botanists studying this flora are constantly finding new species. Even some of the most desirable trade species have not yet been scientifically recognised or described! The intensive harvesting of natural populations by local folk is now a serious threat to the future of the most desirable types. In association with the World Wide Fund for Nature, the University of São Paulo has mounted a major research project and Kew plays a part in this by helping to sort out the classification of the plants.

One of the largest species of pipewort, *Paepalanthus* (Eriocaulaceae), which is common in moist places in the *campo rupestre*, Serra do Cipó, Minas Gerais state.
Photo: S Mayo

The flat-topped mountains (inselberg formations) near Lençois, Chapada Diamantina.
Photo: S Mayo

Work over the past two decades on the *campo rupestre* has revealed that the flora is extremely rich, comparable with that of the Cape in South Africa or the rich rainforests of western Colombia though quite different in every other way. Most of the region is inaccessible, except by foot, and as a result botanical collections are mainly confined to areas along the existing road network. To go further afield, the only practical solution is to use mules, just like the travelling naturalists of former days. On the 1977 expedition, time was too short for such adventures. Our foray began badly, with a car accident and a worrying few days when infected insect bites on Ruth's legs threatened to turn into a general septicaemia.

Our first camp was on the white sandy banks of the Rio Paraguaçú, a tea-coloured river which reaches the sea at Salvador. On the other side of the river from our camp lay a tiny tumbledown village, from which the throb of ritual drums came floating across one fine night. We learned that this was a Candomblé ceremony – essentially an African religion with an admixture of Catholic symbolism – and the following night we were invited over to take part.

The road between Andaraí and the village of Mucugê was our next major collecting site, and proved to be extremely rich. This region lies on the east side of the Chapada and is much wetter; it once supported an extensive area of forest, now mostly cut, along the eastern margin of the mountains. The *campo rupestre* flora is also much lusher. Among many new species collected from the area in recent years is the stemless palm *Syagrus harleyi*, which is common along the Mucugê road. Due to the outstanding interest of its flora and the many recent collections, the area was selected for the first of a series of books on the *campo rupestre* flora being written in association with the University of São Paulo (*Florula of Mucugê* by R M Harley and N A Simmons, 1986).

The aroids posed some problems in this area. A large rosette philodendron (*P. insigne*) was found here and there in shady moist rock crevices and seemed curiously out of place. This is a typical rainforest species of the Guianas and coastal Bahia and not the sort of species one would expect to find on a bare mountainside. Such a find suggests the former presence of much more extensive forests, perhaps in an earlier age when the climate was moister.

From Mucugê we travelled north to the quaintly named town of Xique-Xique (pronounced 'Sheeky-sheeky'), on the banks of one of the great rivers of Brazil, the São Francisco. Our intention was to explore the northern end of the Chapada, and we found it to be quite different: very much drier, and mostly covered with *caatinga*, the deciduous thorn scrub that covers most of the semi-arid interior of northeast Brazil. Our sojourn in this region coincided with Carnival and we were able to see this famous festival in a truly rustic context. Our companions from relatively metropolitan Itabuna commented gloomily that even the songs they sang were 30 years out of date! Ash Wednesday is the day that the men atone for their sins, and while waiting to depart that morning we were assailed by roving bands of women whose aim was to drench us with water, a traditional penance. As I didn't realise what was happening in time, despite seeing a local shin up a telegraph pole to escape his impending fate, I caught a bucketful and paid for it with a bad cold for the following fortnight. Raimundo's assurances that the soaking was a sign of affection did nothing to improve my humour.

Having collected in some very curious *cerrado*-like vegetation (seasonal open woodland typical of central Brazil), which yielded many new species after the specimens were later examined by specialists, we found a substantial lake not far from the Rio São Francisco surrounded by a great grove of carnaúba palms (*Copernicia prunifera*). These plants are typical of damp areas in the *caatinga* and their leaves yield carnaúba wax, still an important export commodity, used for the manufacture of shoe-polish amongst other things.

Climbing the Pico das Almas

We returned to base on 7 March, and immediately began reorganising our collections and restocking with stores for the next and final trip. Our objective now was to visit the southern tip of the Chapada Diamantina, and in particular the Pico da Almas (*c.* 1800 m, 5900 ft). Ray had visited this area in 1974 but had had insufficient time to investigate the area fully.

The nearby towns of Rio de Contas and Livramento da Nossa Senhora do Brumado were classic collecting localities of the German botanist Carl von Martius, initiator of the monumental *Flora Brasiliensis*, and the area was thus of particular interest to us. It may surprise the non-botanist to discover that old collecting sites can be just as interesting as unexplored areas. The reason is that if new species were described from these early localities, the chances are that the species have never been collected again, especially in the tropics. Thus, plants

Heliconia marginata, which is related to the gingers and bananas, in Trinidad, photographed during fieldwork for the *Flora of Trinidad* which was edited at Kew.
Photo: D Philcox

Spectacular towering peaks of Mt Roraima in Guyana. Several Kew expeditions have recently visited this area and a helicopter airlifted some of the botanists on to the El Dorado Plateau.
Photo: P E Brandham

The sundew *Drosera roraimae* on Mt Roraima, showing the sticky droplets that catch insects.
Photo: P E Brandham

A very rare pitcher plant *Heliamphora nutans*, growing in boggy places in the El Dorado Plateau, Mt Roraima.
Photo: P E Brandham

A flowering stem of *Gynerium sagittatum*, a tall grass allied to the European reed, held by T A Cope during a collecting expedition in preparation for the Kew publication *The Grasses of Bolivia*.
Photo: S A Renvoize

Oxen are still valuable for transport in Bolivian lowlands where they graze on the grasses. However, the conversion of tropical forest in Brazil to grassland for beef cattle is environmentally disastrous, as it soon degrades and becomes less productive. The original species-rich forest is replaced by impoverished secondary meadows.
Photo: S A Renvoize

may be known only from a few incomplete scraps and this can lead to all kinds of problems for classification. Re-collection and improvement of our understanding of previously described species is just as important as finding new ones. Contrary to a commonly held belief, taxonomic botany is nothing like stamp collecting – a single specimen is only the barest scrap of information about a species and quite insufficient to understand its relationships and structure properly.

As we began to collect around Rio de Contas we stumbled one day on an old road leading towards Rio de Contas along a small river valley, hidden and overgrown with vegetation. We allowed ourselves to imagine that this must be the same road that poor Martius took 160 years before, in his epic journey across the interior. Just before this section of his journey he and his companions had nearly starved to death and were rescued in the nick of time by another traveller who was taking the same route from the Rio São Francisco to the coast.

Our own journey to the Pico, whilst hardly comparable, gave some problems as the road was very rough in places, apparently no more than a series of flattened boulders. However, with two vehicles we were able to get to the top by towing one another through the worst patches. A fine earthbrick house stood at the roadhead, on the lower slopes of the mountain, and was then still occupied by a dear old widow called Dona Silvina, who at 86 was finding life a little hard on her own. She invited us to camp in her backyard and thus began perhaps the most rewarding week of the whole expedition.

On successive days we roamed ever further up the mountain, finding huge quantities of plants. After two days our drying frames and presses were overwhelmed with material, and Ray was having to write notes at midnight for the previous day's collections. We obviously needed months rather than days to do justice to the area. On the third day we climbed all the way to the summit, and spread out before us were new ranges to the north which are still unexplored by botanists. The number of xyrids (Xyridaceae) and pipeworts (Eriocaulaceae) was

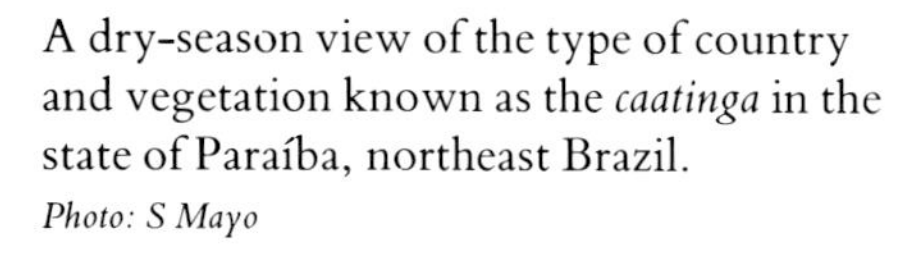

A dry-season view of the type of country and vegetation known as the *caatinga* in the state of Paraíba, northeast Brazil.
Photo: S Mayo

astounding, and virtually all seemed to be different from those we had seen in other *campo rupestre* areas further east and north.

As we collected and discussed our results we began to realise that there were similar patterns of differentiation between the species of many different plants. The mints (Labiatae), aroids and tree lilies (Velloziaceae), for example, all seemed to have endemic species in the same places. The task of working out their precise evolutionary relationships could not of course be started until we returned home, but we could see that such studies were possible and could give exciting new insights on the history and evolution of the plants of the whole mountain habitat. Later work has shown that the picture is extremely complicated and involves links not only within the Brazilian mountains, but between them and those of the Guianas and the Andes. Future studies will look at these problems in particular plant groups.

Our final days in Rio de Contas were memorable ones, of sweltering days and starlit nights on the mountainside, the warmth of the people, the satisfaction of a job well done, and *saudade*, the pain of parting from our stalwart companions. On 31 March, after three months in the field, our thoughts during the long dusty drive back to Itabuna were mainly of how we would organise the next expedition.

In all the expedition collected 2400 'numbers' that is, gatherings of particular species; most collections were in sets of 10 duplicates each, giving a total of approximately 23,500 specimens. The job of naming these and distributing them to specialists and botanical institutes in Brazil and around the world took many years, and is still not completed in those families for which there are no specialists. There are not enough taxonomic botanists to cope with the classification of the world's plant species!

Later developments

Further Kew expeditions followed, with CEPEC and with the University of São Paulo. As I write (November 1988), there is a major expedition in progress on the slopes of the Pico das Almas involving a collaboration between all three institutes. The results of the past 10 years of activity on the flora of the *campo rupestre* and of Bahia have been very encouraging. The CEPEC Herbarium has grown enormously and is now one of the most important in the whole northeastern region of Brazil. Two new national parks have been declared by the federal government in the *campo rupestre*, and the botanical data played an important role in securing these. A series of books on the plants of Bahia and the *campos rupestres* of Minas Gerais and Bahia have appeared, and more are planned. Numerous important postgraduate studies have been completed at the University of São Paulo, based on intensive studies in Minas Gerais. But with all this activity and so many results, the area actually surveyed remains but a tiny fraction of the whole; the scientific potential of the botany has only begun to be tapped.

View of the Sir Joseph Banks Building, Centre for Economic Botany, at the Royal Botanic Gardens, Kew. In the foreground a modern armillary sphere depicting astrological herbs for each month (designed by Brookbrae Ltd), and the Herbarium in the background.
Photo: A. McRobb

PART III

RECENT KEW SPECIALIST EXPEDITIONS

11 *Quest for useful legumes*

GWILYM LEWIS

Kew has a long history of involvement in the study and introduction of useful plants. In 1848 Sir William Hooker opened the world's first Museum of Economic Botany at Kew, while 1989 saw the completion of the Sir Joseph Banks Building, Centre for Economic Botany, beside the Herbarium. This centre houses hundreds of thousands of samples of plant products from timber to textiles and reeds to resins, indexed on computer. Another database holds information on plants used in arid and semi-arid lands. We have already recalled in Chapter 10 of this book the famous introduction of rubber; early issues of *Kew Bulletin* recorded information on many other plants of economic importance. Today the search goes on with even greater intensity and, as an example of recent studies and fieldwork, the legume family has been chosen here as one group of world importance.

The well-known pea and bean or legume family (Leguminosae) is the third largest flowering plant family after the Orchidaceae and Compositae, comprising about 670 genera and 17,500 species worldwide. The Leguminosae form the major component of many vegetation types around the world. Habitat preferences vary from mountain top to coastal sand and from tropical rainforest to desert; there are even a few aquatic legumes. Many species in the larger genera are highly characteristic of open and disturbed places. They are eminently suited to early colonisation and subsequent rapid expansion in such habitats due, in a large part, to their associations with either nitrogen-fixing bacteria or with root fungi. Bacteria of the genus *Rhizobium*, housed in the root nodules found on many legume species, are able to convert nitrogen gas from the air into ammonia (a soluble form of nitrogen accessible to other plants) and these legumes are extremely valuable as natural fertilisers.

Economic uses of legumes

Many of the beautiful and characteristic street trees of the tropics are legumes – *Delonix regia, Caesalpinia peltophoroides, C. pulcherrima* and *Erythrina crista-galli* to list only a small sample. The seeds, pods, leaves, roots and flowers of many legumes provide a protein-rich food source for humans and animals in nearly all parts of the world, and the large-scale cultivation of some species produces valuable cash-crops.

Some of the finest timbers come from legumes; various rosewoods from *Dalbergia* species are highly valued for cabinetwork and carving and the best-quality violin bows are still made from *Caesalpinia echinata*. Some of the best copals – viscous resins – widely used in the production of varnishes, paints and lacquers are produced from leguminous trees, particularly of the genera *Hymenaea* and *Copaifera*; others are a source of tannin for the leather industry. Rotenone, an insecticide and fish poison, is extracted from species of *Derris, Lonchocarpus* and *Tephrosia*.

Many processed foods contain leguminous gums; several legumes – *Hymenaea courbaril, Gliricidia sepium* and many species of *Andira, Acacia, Inga* and *Albizia* for example – are prized by beekeepers for producing some of the finest honey. Legumes are used to flavour pickles, curries and other foods and to produce liquorice, perfumes and oils. They also produce some important dyes, such as the brilliant blue 'indigo' from some *Indigofera* species.

Legumes occur in the production of medicines; as a component of oral contraceptives (several species from the South American genus *Brownea* for example); and in some instances as hallucinogenic drugs. Both species of the neotropical genus *Anadenanthera* have been widely used in the preparation of a hallucinogenic snuff by Amazonian Indians.

Pods of *Entada polystachya*, a robust leguminous liana with fruits breaking up into one-seeded, wind-dispersed papery 'envelopes'.
Photo: G P Lewis

Specialised collecting according to plant families produces high-quality specimens of increased scientific value. In the 1980s more botanists are spending time in the field, studying floral biology, pollination, breeding systems and flowering periods (phenology). There is a real practical need to study plants *in situ* and in detail. We also need to study tribal peoples and the uses to which they put their regional flora, and we should not just extract the information and then forget the source – the people and their environment need to be protected as an inseparable unit.

Kew botanists are often asked whether it is commonplace to discover new species on expeditions. They always answer yes, because particularly in the tropics the flora is so rich and so little is known about it. In Bahia alone (one of the 23 states of Brazil and approximately the size of France) in 1987 I catalogued 741 species of legume, 92 of which had been described as new to science since 1979. Our current knowledge of the South American flora is very limited compared with data available for Africa. Over the past 15 years there has, however, been an increase in the number of Kew Herbarium staff researching plant groups from the New World tropics, particularly Brazil, and this has resulted in a proliferation of generic revisions, monographs, checklists and floras.

Useful plants and plant products may be seen on sale all over the world, such as at the Sunday market at Chinchero, near Cuzco, Peru. Held close to the ancient Inca ruins, this is a traditional market for local fruit and vegetables, and also handicrafts of plant products for the tourist trade. Among the food plants on sale were seeds of the lupine *Tarwi* (*Lupinus mutabilis*) and other legumes, various cultivated forms of potato (*Solanum tuberosum*), which is native to this region, some high-altitude tuber crops such as *Olluco* (*Ullucus tuberosus*) and important seed-crops such as *Quinoa* (*Chenopodium quinoa*).
Photo: P Rudall

Brazil and brazil-wood

Since 1978 I have concentrated my studies on the legumes of Brazil and two of my latest research programmes have centred around *Caesalpinia* (a genus occurring throughout the Tropics with many species endemic to Central and South America) and the preparation of a checklist of the legumes of an inland Amazonian island, called Maracá.

Of historical interest is the fact that Brazil (spelt 'Brasil' in Portuguese) was named after a local species of *Caesalpinia* – *C. echinata*, commonly called *pau brasil, pau pernambuco* or 'brazil-wood'. But there is more to the story than that. As early as the 11th century timber from *Caesalpinia sappan*, a sister species of *C. echinata*, was being exported from Asia to Iran as the source of a wine-red dye called *bakham* by the Iranians. The dye was carried to Europe, where it was used for tinting woollens, cottons and silks red. *Bakham*, translated into Latin, became *brasilium* and the original brazil-wood thus originated in Asia.

As early as 1193 in the custom houses of Ferrara, Italy, notices appeared of a dye for colouring woven goods red called Brazil, Brecillis, Bracire, Brasilly, Brazilis and Brazili, which the Italians thought derived from *brasa* or *braise* (meaning red) because of the red colour of the wood. When the explorer Pedro Cabral set foot on the east coast of South America in 1500, he celebrated his discovery by holding Mass on the beach and christening the new territory the 'Land of Santa Cruz'. Portuguese settlers subsequently discovered forests of

Cassia moschata, an elegant leguminous tree growing near the base camp on Ilha de Maracá. Many species of *Cassia* (and its segregated genera *Senna* and *Chamaecrista*) are of economic value, being used for medicines, vermifuge, tannin, timber, firewood and other purposes.
Photo: G P Lewis

Caesalpinia echinata producing similar timber and dye to that from *C. sappan* and Santa Cruz was ultimately renamed 'Brasil'.

Brazil-wood became a generic term applicable to a group of red or orange-coloured dyewoods derived from several species of closely related caesalpiniaceous genera of pantropical distribution. The only American species of commercial importance for its timber remained, however, *C. echinata* with the best-quality wood being found in the east Brazilian state of Pernambuco, hence its designation as pernambuco wood. All Brazilian brazil-wood was at one time in such great demand in Europe that the business was made a Portuguese royal monopoly in 1623, and private exploitation was strictly prohibited until the middle of the 19th century. Today a few cottage industries in Britain still use the dye from *C. echinata*, but the principal foreign demand is not for dyes but for the manufacture of violin bows. Professional violinists will not consider using a bow made from anything other than pernambuco wood and high-quality bows now sell for as much as £1000.

After an extensive search of several Brazilian, European and North American herbaria I was able to conclude that very little botanical information was available about *Caesalpinia echinata* and herbarium specimens have little information about the plant's habitat preferences – most simply cite the collection locality as Brazil, some adding 'Atlantic forest' and the name of the state.

Looking for brazil-wood trees

In May 1987 I set out with colleagues from the botanic gardens of Rio de Janeiro to look for *Caesalpinia echinata* in the coastal forests of Brazil. Our aims were to discover where the species was still growing, whether it was endangered in its natural habitat and, if so, to decide what measures were needed to save the species from extinction. In addition we wanted to look at variation in form, phenology and seed-development. Our collecting trip was to take us up the coast from Rio de Janeiro to Porto Seguro in southern Bahia, and although concentrating on the study of one species we also planned to collect other interesting legumes.

Provided with all the equipment necessary for a botanical collecting trip, we left Rio expecting to discover the worst – that *Caesalpinia echinata* was already near extinction – so it was a pleasant surprise to find the species surviving within a few hours' drive from Rio. Even more surprising was the type of habitat in which the plant was discovered and that the tree was not a large forest giant but a small, contorted 5 m (16 ft) treelet. It was growing amongst coastal cactus scrub on low boulder-strewn hillocks within a few hundred metres of the sea. There had been no indication of this habitat preference on herbarium labels and it seemed likely that we were looking at one form of a variable species or perhaps even an undescribed one. Nevertheless the bark was armed with scattered spine-tipped warty protuberances and the fruit was spiny like a small hedgehog, both characteristics of the species.

The Brazilian tree *Caesalpinia echinata* is known as *pau brasil* or 'brazil-wood' and gave its name to the country because of its valuable timber and red dye sap. It grows in the coastal forests, which are diminishing rapidly.

Drawn by Tim Galloway

A few days later, further north in the state of Espirito Santo, a local forest guide was able to take us to fine 40-m (130 ft) tall specimens of *C. echinata* with trunks 1 m (3 ft) in diameter. These produced copious wine-red sap when cut with a machete. They were growing in quite a different vegetation type – true Atlantic coastal forest – and their spreading crowns emerged above the general forest canopy. I could not help wondering how many £1000 violin bows could be produced from each tree. After a more careful search of the forest, we realised that the total number of these trees was very small and we were studying only a small remnant of what was once a much greater forest zone.

This became the norm as we travelled northwards – we were able to find one or two specimens in remote forest pockets but the species was never abundant. In the more accessible areas, nearer the coast, most trees had been cut down and only regrowth side shoots were evident. The locals use the smaller tree trunks for charcoal production and even as fence posts, and some areas of coastal sands, where the species once grew, are now rubbish tips or have been razed to the ground to permit the building of holiday homes. In addition, huge areas of Atlantic coastal forest have been cut down and replaced with crops of sugar cane and cocoa and plantations of fast-growing *Eucalyptus* species. Information from the violin-bow industry in Europe indicates that European timber traders are still importing the species from Brazil, further threatening the plant in the wild.

However, the situation is somewhat alleviated by the successful cultivation of *Caesalpinia echinata* in at least two forest reserves in Brazil – one in Porto Seguro, Bahia, and the other in Linhares, Espirito Santo. The species is also being planted in many cities as an elegant ornamental tree and there are some attempts to re-introduce it into the wild. It is equally important to protect the species with its associated pollinators

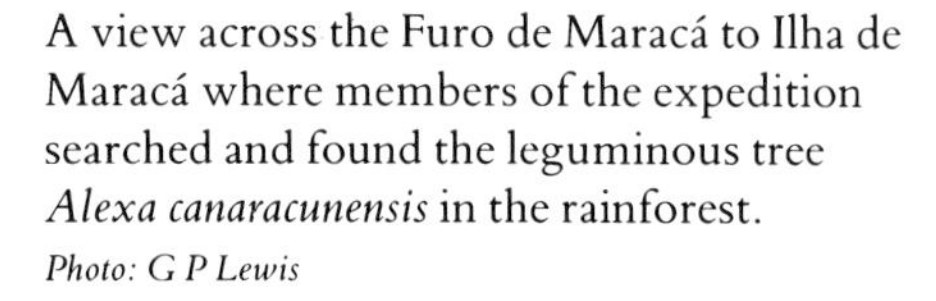

A view across the Furo de Maracá to Ilha de Maracá where members of the expedition searched and found the leguminous tree *Alexa canaracunensis* in the rainforest.
Photo: G P Lewis

Brazil is well endowed with cacti, such as *Melocactus smithii* (foreground) and *Cereus hexagonus* (mid-distance) seen here growing on a granite outcrop at Pedra de Wanawana, Ilha de Maracá. The natural populations of cacti have been much damaged by large-scale commercial collecting, which is now being controlled by international legislation and export controls.
Photo: G P Lewis

in their natural habitat, to encourage the locals to use other species for the production of charcoal and to attempt to discover acceptable alternative woods from which to make violin bows. Many Brazilians had assumed that their national tree was already extinct. The good news is that it is not, the bad news is that, in the wild, it soon could be. Urgent conservation measures are required to prevent the loss of such an historically important species from the coastal forests of eastern Brazil.

The Maracá Rainforest Project

At the end of 1983 the Royal Geographical Society (RGS) of London was invited to work on the Amazonian island of Maracá by Dr Paulo Nogueira Neto, the then secretary of the environment in Brazil and director of the Secretária Especial do Meio Ambiente (SEMA). Dr Nogueira asked the RGS to undertake a general survey of the island's vegetation, fauna, geology, geomorphology, soils, hydrology and climate. This invitation was accepted and five research programmes were formulated with the main aim of providing detailed information about the island's complex tropical ecosystem. One of these five programmes was an ecological survey of the island, and the RGS invited Kew to send a team of botanists to prepare a botanical inventory of Maracá.

The Maracá Rainforest Project was officially launched by its patron, HRH the Prince of Wales, on 4 December 1986 at the Royal Geographical Society in London. The main fieldwork phase lasted from February 1987 until March 1988. Kew's participation was designated its main fieldwork commitment in the tropics in 1987, and it sent four plant collectors to the island. Each of the botanists

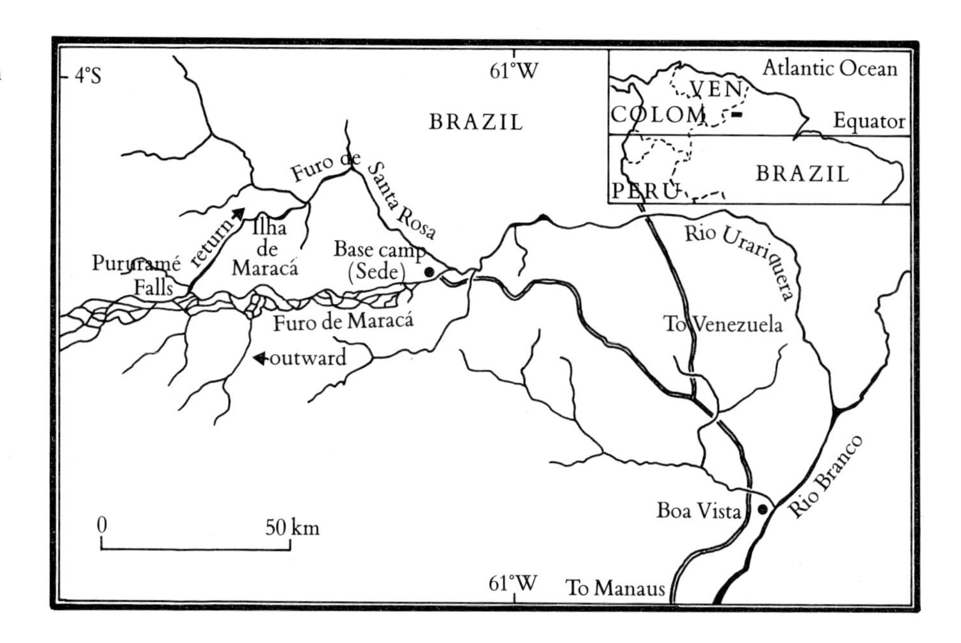

Map 13 Location of Ilha de Maraca in north-east Brazil, with places mentioned in ch. 11.

concentrated on collecting plants in a specialist group, namely Leguminosae, Loranthaceae, Labiatae and Pteridophyta: I concentrated on the legumes. In addition all four made general collections as part of the flora survey of the island.

The Ilha de Maracá is one of the world's largest riverine islands, covering an area of some 92,000 ha. (*c.* 227,000 acres). At the western tip of the island the Rio Urariquera divides to become the Furo de Maracá to the south and the less treacherous Furo de Santa Rosa to the north, rejoining at the eastern end of the island. Apart from the family manning a small fieldstation at the eastern end, Maracá is uninhabited. With its conveniently defined boundary and lack of disturbance by man, it provides an ideal area for innovative biological research.

Most of Maracá is fairly flat, although in the north and west there are some low peaks reaching 200–300 m (650–980 ft) above sea level, and to the south there are some fascinating granitic outcrops that support at least two Amazonian species of cactus (*Melocactus smithii* and *Cereus hexagonus.*) The river channels surrounding the island are up to 50 m (160 ft) wide and there are frequent, scarcely navigable rapids on each, particularly numerous on the Furo de Maracá. The underlying geology is of Lower Precambrian age, being part of the ancient Guianan Shield, but not far to the north are the Roraima–Pacaraima sandstone escarpments that form the watershed between the Amazon and Orinoco basins. This area, known as Pan-tepuia, is characterised by flat-topped mountains, known as *tepuis*, including Roraima itself. The nearest *tepui* to Maracá is the small and isolated Tepequem, some 30 km (19 miles) from the northernmost point of the island.

I made three trips to Maracá, the first from February to April 1987, the second from October to November the same year and lastly a short visit in February 1988. From June to September 1987 I was back at Kew and able to identify the dried material collected during the first visit to the island. It became clear that Maracá is very rich in legume species.

AIDS research and the legumes

During the same period the legume department at Kew was very excited by the results of some research jointly undertaken by the Jodrell Laboratory at Kew and St Mary's Hospital in London. It was demonstrated that the chemical castanospermine, isolated from the Australian endemic tree *Castanospermum australe*, was giving encouraging results in *in vitro* trials against the AIDS virus. A colleague, Charles Stirton, was engaged on a taxonomic revision of neotropical species of the leguminous tribe Sophoreae, to which *Castanospermum* belongs and the Jodrell staff asked if he could suggest closely related genera in the neotropics that might have a similar chemical composition. Charles suggested the genus *Alexa* as one possible candidate and Rob Nash set to work at the Jodrell running samples of seed and leaf material to see if castanospermine or closely related compounds might be found. The results were positive. Large samples were now needed, preferably from freshly collected material and quickly.

Charles knew that I was about to return to Brazil and that Maracá was within the area of distribution of *Alexa* species – there are 10 species in the genus distributed from Venezuela through the Guianas and into the northern part of the Amazon basin. A careful search of the herbarium material at Kew pinpointed one species collected at Tepequem and it was decided that I should attempt to collect material of this species – *Alexa canaracunensis*. If this meant securing a helicopter flight from the nearest town, Boa Vista, then no expense should be spared! I familiarised myself with the morphological characteristics of the plant and its preferred habitat and prepared photographs of herbarium specimens to take with me as a reference source.

Return to Maracá

I arrived back on Maracá on 10 October after a flight from Rio northwards to Manaus, a connecting flight to Boa Vista, a three-hour jeep ride to the River Urariquera and a short boat-trip across the Furo de Maracá. I was to continue to collect legumes, hopefully adding to the checklist by collecting at a different time of year and, if possible, by camping out at the western end of the island. Much of the island was now accessible by boat and the rivers were high enough to allow clear passage through most of the rapids.

The first person that I met on arriving back at base camp was William Milliken, an ex-Cambridge University student and the only botanist working full time on Maracá. William had been very active since I left in April, cutting new trails all over the island, setting up sub-camps and collecting many of the large forest trees as they came into flower or fruit. He was keen to show me some of his more interesting legume collections and I was more than happy to spend a pleasant hour or two attempting to identify these gatherings.

I could hardly believe my luck when he flipped open the newspaper protecting the first specimen to expose a well-preserved example of *Alexa canaracunensis*. Where was it from? What did William know about the tree? When did he collect it? Again luck was with me – the tree was a dominant species in the forest at the western end of the island and William had collected the sample I was looking at only 10 days earlier after a boat journey up the Furo de Santa Rosa. I explained the significance of the find and determined to get to the other side of the island by any means available. William had been considering a trip up the more dangerous Furo de Maracá and particularly wanted to map its complex system of rapids, and four Indians from the Maiongong tribe near the Venezuelan border had just been employed by the expedition and were highly acclaimed as excellent boatmen. Our plans fell together with remarkable ease and it was decided that William and I, together with three of the Indians, should attempt a circumnavigation of the island. I confess to a certain nervousness about the trip, largely resulting from having been in an unpleasant boating accident on the River Urariquera in April of the same year. (On that occasion I and several companions had lost all our camera equipment and some valuable collections.)

William had witnessed the power of the Pururamé Falls at the western end of the island and his observations fully supported those of earlier explorers – a boat could not travel through the falls but would have to be carried for one kilometre around them. With this uppermost in our minds, we restricted our collecting and camping equipment to the minimum and limited our food supply to the bare essentials for five men for six days. We would supplement our diet with fish from the Furo de Maracá and fruits from the forest. Fuel for the 25 hp Mercury engine was an essential commodity – we would have to rely on paddles if we ran out of petrol.

We all knew that the journey was not going to be an easy one and prepared our equipment for a rough ride; cameras, food and clothes were packed into thick plastic bags; boxes, petrol cans and collecting equipment were buckled down to the floor and seat supports of the small, aluminium craft, and the whole baggage area was covered by a thick plastic sheet. A small shovel, several machetes, two paddles, fishing tackle and one lifejacket – mine – concluded the inventory and we were ready to leave.

Upriver to the Pururamé encampment

On the morning of 16 October we set off – two tall, thin, pale-skinned Englishmen and three well-muscled, stocky bronzed Indians. How amusing William and I must have seemed to them. The morning progressed peacefully as we sped upriver. There was little necessity to jump out of the boat to haul it through the rapids for the water level was still high and Luiz, our boatman, always knew which of the many channels was the best one to navigate. At the more dangerous points he

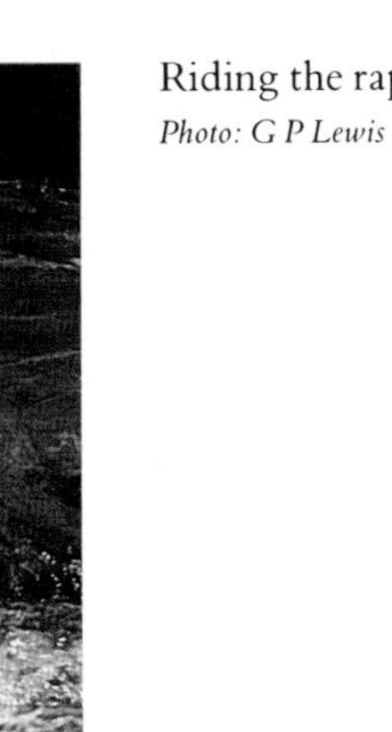

Riding the rapids of the Furo de Maracá.
Photo: G P Lewis

would always motor up to the lip of the rapid, where the water thundered over the protruding boulders, and consider the possible routes before allowing the boat to drift back downstream for ten metres or so. Without further consideration, he then opened up the throttle and we raced upriver; water crashed over the bows and filled the bottom of the boat and we bailed frantically to keep afloat.

Eventually, however, we did have to leave the boat and struggle through the fast currents, dragging the boat over the slimy rocks and broken tree trunks. This is when I abandoned the lifejacket, for it is quite impossible to move upstream in a fast-flowing river, scrambling over slippery rocks in chest-high water wearing a jacket designed to keep one afloat. Floating horizontally, hanging on to the boat for fear of being swept downriver did not help to the team effort of trying to move the boat forwards! As the day progressed, teamwork improved considerably – I knew what I could achieve and also when I was being more of a hindrance than a help.

Occasionally we had to cut away the tangled riverine forest above our heads or saw through submerged tree trunks blocking the narrow channels. We disturbed beautiful birds sunning themselves on the riverbanks or on rocks midstream – egrets, herons and diving birds sped away as we approached. Giant Amazonian otters – now an endangered species – were a rare treat, seen bobbing up and down as they sped along in the river currents, calling out to each other like happy children at play. Water snakes, capybara, tapir and alligators were more common and we were lucky to see pink river dolphins.

At a quarter to six, just as the evening light was fading, we paddled down a narrow side-channel, moored the boat and set about preparing camp. Within 20 minutes the plastic sheet was lashed over a central pole and secured to four trees, thus providing an ideal shelter. The hammocks and mosquito nets were secured and a camp fire flickered in the

Wildlife other than plants is often encountered during expeditions. In Brazil during the Ilha de Maracá trip this iguana (*Iguana iguana iguana*) and the arrow poison frogs (*Dendrobates leucomelas*, also illustrated) were encountered. The maintenance of natural habitats depends on an intricate balance of animals and plants still little-understood.
Photo: G P Lewis

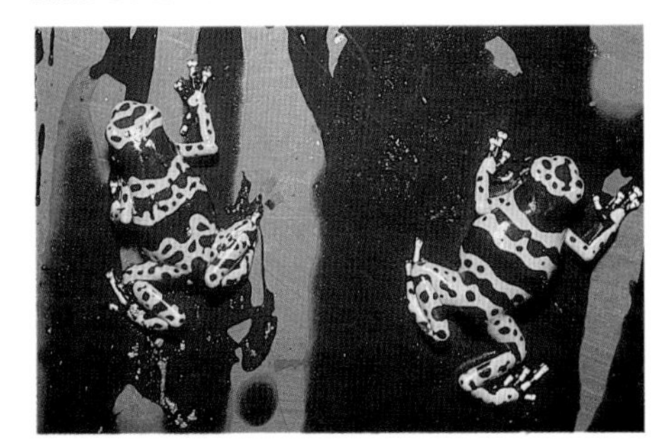

darkness. The sounds of the night animals competed with the rushing river as we settled down to sleep early. As I dozed off I had time to reflect upon the day's journey and what a privilege it was to be travelling with such an experienced crew.

The following day we got up with the morning light and were soon under way. Our progress became a familiar routine and this allowed more time for observing the riverine forest. Occasionally we stopped to collect from the tangled mass of vegetation festooning the river-banks; *Swartzia grandiflora, Acosmium tomentellum* and *Ormosia smithii* were three leguminous trees collected as we continued upriver. By early evening we had reached the western tip of the island and now turned northwards on the Furo de Santa Rosa, reaching the Pururamé falls by five in the afternoon.

We decided to camp above the waterfall and found an old gold-mining encampment which provided the basic framework. The Indians then busied themselves fishing in the shallows. Rancid chicken was used to catch the ferocious piranha, and fruit of a small palm (*Bactris maraja*) proved to be an irresistible bait for the pacú – a species related to the piranha, but less bony. We were up again at the first sounds of dawn and had soon breakfasted and broken camp. Within a few minutes we had navigated a series of rapids above the falls and beached the boat where a narrow Indian trail traversed the hilly ground, skirting the main waterfall. The boat was emptied of all its contents and we decided to carry it above our heads the kilometre to the base of the falls. This was a mistake: very soon our arms were being wrenched from their sockets and movement forwards was hindered by the difference in our heights. We resolved to build a sled and to push and haul the boat uphill and allow it to slide downhill. The final short stretch was down a 60-degree incline and little more than a mud slide. We arrived on our bottoms at high speed!

We had reached our goal, the Pururamé encampment, which is in a beautiful location on the bank opposite the island. Many earlier explorers had camped here, as witnessed by the names carved in the rocks – General Rondon, Hamilton Rice and 'S' for Schomburgk. And the early explorers had generously considered future travellers; around the camp site they had planted lemon trees and bananas and these were now heavy with fruit. A huge *Swartzia* shaded the site and by the river an old specimen of *Macrolobium acaciifolium* served as an ideal mooring for the boat. Two further trips along the slippery path were needed to bring all the other equipment round to the camp site. Marco, the youngest and smallest of the Maiongong, carried the heavy outboard engine on his shoulder with ease – a demonstration of the remarkable strength and fitness of these people.

Finding *Alexa*

Having set up camp and lunched on fish, rice and lemons, William and I set off to collect around the camp while the Indians went fishing. Within 20 minutes we had found a specimen of *Alexa canaracunensis*

The quarry of the expedition was a leguminous tree known locally as *Tunadi* and scientifically as *Alexa canaracunensis*. Its flowers (a) develop into pods (b), and the seeds as well as leaves yield compounds that may act against the AIDS virus.
Photos: W Milliken and G P Lewis

and, as previously confirmed by William, the cut bark gave off a strong odour of freshly-sliced cucumber. The tree trunk was about 10 cm (4 in) in diameter with fissured greyish white bark and an elegant crown of large pinnate leaves. While I took photographs and made notes about the habitat, William climbed the tree and in under one minute was swaying about in its crown.

He cut one or two inflorescences (flower-heads) and some leaves and then let out a yell followed by a few well-chosen words and was on his way down again as fast as he could go. The younger branchlets have hollow stems and in these live colonies of little black ants – as soon as William had cut the first branchlet they had swarmed all over him. This was a new discovery – unreported in the literature – and we quickly collected a sample of the ants (now known to be a species of *Procryptocerus*). A second exciting discovery was that each flower has a pear-shaped fleshy bract which supports several pinhead-sized glands – these glands had not been apparent on herbarium material. Similar glands were found between some pairs of leaflets and we concluded that the ants probably use them as a sugary food-source. It is probable that in return for providing a home and food supply for the ants, the tree is protected from potential herbivores by the ants as they patrol the tree in search of the glands; the lack of insect-damaged leaflets further supported this hypothesis. We collected herbarium material of flowers, fruit and foliage and samples of bark, wood and root tips. Flowers were fixed in alcohol for subsequent anatomical study and I used one roll of film photographing every aspect of the plant. The creamy white petals of the flowers contrast beautifully with the chocolate-brown hairs of the calyces and fruits.

The following day we arose before light, keen to have a full day's collecting on the island. The day before, Luiz had cut the sole of his foot deeply on some sharp palm spines and as I returned from bathing in the river he was to be seen sitting cross-legged under his hammock, stitching together his wound with a large needle and some bright green cotton. He closed the cut with ten perfect cross-stitches and admitted that the self-administered operation *was* rather painful!

Drying herbarium specimens is a laborious task: Kew botanist Peter Edwards arranging the plants between absorbent paper and Raymond Harley writing notes during the Ilha de Maracá expedition in 1987.
Photo: G P Lewis

William and I explained that the main purpose of the day was to cross the river back to the island and collect as much *Alexa* fruit as possible. On the island we climbed up through the dense riverine forest into a more open forest at about 100 m (320 ft) altitude. Here the understorey was rich in *Musaceae, Rubiaceae, Heliconiaceae* and palms with the legume *Parkia pendula* the largest emergent tree, attaining 40 m (131 ft). We followed the trail cut some 10 days earlier by William's team. It was evident that *Alexa canaracunensis* was one of the dominant species in this hill-slope terra firme forest and we set about collecting more material – an arduous task as each tree only supports one or two inflorescences and each of these only one to three mature fruits. Investigative bark slashes made on the previous trip were now coated in a protective jelly-like exudate and we collected samples of this for future analysis. Similar exudates were observed on other tree species so these were collected, together with herbarium specimens.

Co-dominant in the forest was another leguminous tree, *Clathrotropis macrocarpa*, a more robust species than the *Alexa*, but with a similar cucumber odour when cut and a similar exudate from the trunk. This species was in full fruit and each tree supported 30–40 large orangish pods, each containing one or two fleshy seeds – one of the largest seeds in the legume family. A bulky sample of these seeds was bagged as I knew that this plant was closely related to *Alexa*, in the same legume tribe *Sophoreae*. There was a good chance that it too might contain castanospermine.

The forest was evidently rich in legumes. Species of *Ormosia, Albizia* and *Inga* were common and the robust woody liana *Bauhinia guianensis* was very conspicuous. By the time we stopped for a streamside lunch-break in an idyllic setting, the three Indians had collected 40 fruits of *A. canaracunensis*. We were all well pleased.

In the evening, discussions revealed that the Indians knew of no local use for *Alexa canaracunensis* although it did have a common name – 'Tunadi'. This little-known species of no apparent local use might prove to be one of the most economically important tree species in the tropics, a fact which supports the conservationist view that we must protect natural habitats for they contain plants of immense potential to man, and as yet we know so little about them.

Subsequent development

Weeks later, back at Kew, staff of the Jodrell Laboratory isolated castanospermine from the bark, resin, fruits and leaves of the *Alexa* species and closely related, as yet unidentified, compounds have been found in *Clathrotropis macrocarpa*. The high levels of castanospermine in the slash exudate present interesting possibilities for the future – perhaps the species can be tapped, as in rubber trees, to provide a constant flow of a castanospermine-rich jelly. But this might not be necessary as it is now clear that the compound can be relatively easily synthesised.

However successful castanospermine proves to be in the fight against AIDS, it is clear that plants could provide many of the future medicines and foods for man – as long as the forests are not destroyed before we can collect and study the species that they contain.

12 New horticultural plants

JOHN LONSDALE

Many plants collected by the great explorers of the past seem now to be essential components of the British garden scene, so much so that it is difficult to believe that many have been known for little more than a century. The great names of George Forrest, E H Wilson, Robert Fortune, David Douglas and many more live on in the plants they introduced during the heyday of plant collecting in the 19th and early 20th centuries, but it would be a great mistake to think that these days are over. Changing political attitudes enable collections to be made from previously closed areas. The speed and ready availability of modern transport also aids the collector, helping to ensure that living material can be successfully brought into cultivation. Although the former flood of introductions is now a mere trickle, new plants are discovered every year. Not all, of course, are suitable garden subjects but some are. Of more certain value are those plants lost from cultivation which are re-introduced by today's collectors, or those which exhibit improved qualities of flower, form or hardiness.

Every year certain introductions to the collections at Kew are found to have particular qualities which make them very desirable garden plants. The Royal Botanic Gardens have a responsibility for the evaluation and distribution of such material to ensure it becomes widely available, thereby furthering its chances of survival in cultivation.

Introducing new woody plants

Trees and shrubs are particularly important to collectors of horticultural novelties. Most of the trees and shrubs found in British gardens originate from other parts of the world. We have very few truly native trees, and fewer still are attractive additions to the small garden. The great collectors of the past have enriched our gardens with magnolias, rhododendrons, camellias, acers, a vast array of conifers and many many more. The result is that to look into any major British garden today is to look at the flora of the entire temperate world.

The present curator at Kew, John Simmons, has with others introduced a number of improved forms of hardy tree species. *Alnus subcordata*, a native of the Caucasus and Iran, has been grown at Kew for many years. It is an attractive tree with deep green heart-shaped leaves and long male catkins in early spring. A form from a new location in the Elburz Mountains of Iran collected in 1977 by Simmons and Hans Fliegner is a much improved selection. In the wild it reached 45 m (147 ft), but it is unlikely to attain this height in Great Britain. It is very vigorous in cultivation and is being evaluated as a potential timber-producing tree for wet sites. Its ability to grow in poor, wet conditions is assisted by nitrogen-fixing root nodules containing a fungus which supplies nitrogen to the trees. This feature makes it of great value for use in land reclamation on nutrient-deficient soils.

Many other trees and shrubs were collected on the same expedition, and most of them were hardy. Some could prove to have an outstanding feature – indeed one, *Parrotia persica*, the Persian ironwood, does seem to hold great promise for the smaller garden, being upright rather than having the spreading habit of the form most often seen in cultivation. Often far wider than it is high, it can (exceptionally) reach 9 m (30 ft), which is a great inconvenience. It will be a great reward if this fastigiate selection results in the plant becoming more widely known, as it is one of the most versatile small trees, having desirable qualities throughout the year.

Even in the dark winter months the tree provides interest with most attractive flaking bark: creamy white young bark contrasts strongly with patches of grey and warm brown, which are in turn complemented by the almost black older bark. In late winter the tree carries clusters of flowers which would be insignificant were it not for prominent deep crimson stamens that suffuse the whole tree with a warm glow. In summer the tree bears bright clean deep-green foliage, which as autumn approaches gradually changes to gold, yellow, crimson and red. It is widely acknowledged to be one of the finest trees for autumn colour.

Another tree worthy of introduction into the smaller garden which has yet to gain recognition is *Sorbus microphylla*. It is slow growing, never reaches a large size and is attractive in foliage and flower. It bears small, toothed deep-green leaflets which are carried in 10–15 pairs along a central stalk. The pink flowers are followed by large white or pink-tinged globular fruits. The tree is a native of the Himlayas and is found from Nepal to China. It has been collected several times by Kew personnel. Tony Schilling, deputy curator at Wakehurst Place, (in 1968) and Brian Halliwell (in 1970) have both brought back seed from trees growing in high-altitude rhododendron–birch forest in the Langtang valley in Nepal. A form collected by Tony Schilling in the Everest National Park in 1978 has white fruits flushed with crimson.

Former Kew student Jim Priest, who visited Ecuador on comple-

Another plant collected in Iran by Simmons and Fliegner was *Parrotia persica*. This species is well known in British gardens where it forms a small tree with a simple trunk, so it came as a surprise to see the massive fluted bole of wild trees (a). Progeny established at Kew have brilliantly coloured leaves in the autumn (b).
Photo: J B E Simmons

tion of his Kew Diploma course in 1985, brought back one of the most horticulturally exciting plants collected in recent years. *Fuchsia loxensis*, from Imbabura Province on the borders of Ecuador and Columbia, is a small plant which produces flowers of clear vermilion, without a trace of the blue pigment which cools the red of most fuchsias. The sepals of individual blooms form a star from which the dazzling petals emerge. Neat rounded foliage of the deepest green further heightens the brilliant impact of the flowers. Seed was collected at an altitude of 3450 m (11,200 ft), and plants from this altitude can be expected to adapt to cold because temperatures can drop to near freezing at night even though the region is tropical. There is a good chance that the Imbabura form of *F. loxensis* will prove to be hardy. It has already shown promise in a sheltered spot at Kew.

The importance of knowing the provenance of any plant intended for outdoor cultivation cannot be stressed too highly. Many plants previously thought to be too tender to be grown in the open have been found to be hardy, providing seed is taken from an appropriate geographic location. Some plants have a very wide distribution in the wild and may range from tropical to temperate regions, others may be more localised but grow at greatly varying altitudes. It is usually those from the highest altitude which are of most value in British conditions.

Daphne bholua (far right) photographed in Nepal, showing a well-grown bush covered in sweetly-scented flowers during early spring. Various shades of pink to almost white occur in the wild and it is expected that hardy forms will become popular in British gardens.
Photo: A D Schilling

Daphne bholua, for example, is a very attractive winter and early spring-flowering shrub, not unlike the common *Daphne mezereum* but grander in every way, being taller and carrying enormous crops of large sweet-scented flowers in shades of pink. The plant has been grown in Britain since the 1930s but has never become popular, although it has aroused more interest in recent years. In its native Himalaya it grows at a wide range of altitudes. Plants from lower

regions are evergreen but those from higher altitudes are deciduous. It is only the latter forms which can be grown reliably outdoors. In the 1980s Tony Schilling has discovered exciting deep-coloured variants of these hardy forms which will hopefully extend the colour range.

During an expedition to South Korea in 1982 Kew horticulturists and members of the Forest Research Institute at Seoul collected seeds of trees and shrubs in the Gaeryong National Park, shown here, and Sorak Mountains.
Photo: C M Erskine

'Award of Merit' for Kew introductions

One method of giving a horticultural 'seal of approval' to a plant is to place it before a panel of specialists at the regular Royal Horticultural Society shows which are held at the agricultural halls, Vincent Square, London. After a period of successful cultivation, plants which are judged to have value and interest to the wider horticultural world are submitted for consideration. Several grades of award are given, any one of which is an indication that the plant has potential and is likely to be worthy of general cultivation. The collections at Kew are continually being assessed for plants which have these qualities. Those that are selected often form the foundation stock for commercial nurseries to multiply and make available to all. An Award of Merit (AM) is an indication that a plant has particular garden-worthy qualities. Many plants derived from recent expeditions hold this distinction.

Rosularia sempervivum, a member of the family Crassulaceae, was awarded an AM in 1985. This family has provided a great variety of ornamental succulent plants – the most widely-grown hardy genus being *Sempervivum*, the familiar house leek. *Rosularia* is similar but not nearly so well known. Plants of this species were collected by Simmons and Fliegner in the Elburz mountains of Iran in 1977. The plant is best grown where it can be protected from excess moisture in winter, when it will repay the effort with masses of tubular pink flowers which completely obscure the foliage.

Antirrhinum molle is another example of a plant collected by Kew staff (Hans Fliegner and Roger Howard) which has been given an award. A relative of the familiar bedding snapdragon, this interesting addition to the rock garden is native to the mountains of the Pyrenees and northeast Spain. The plant has a sprawling habit and carries creamy-yellow flowers tinged with pink or red which fade to white as they open. The leaves are covered with soft grey hairs.

South-American specialities

A few of my own Argentinian collections have proved to be reliable and worthwhile plants. It is not always the most striking plants which turn out to be the winners in cultivation. Often plants which take the eye for no obvious reason can far exceed expectations and an eye for the unusual can pay unexpected dividends.

In 1978, collecting in a remote part of Misiones Province in northern Argentina (map, p. 178). In a dense thicket of stunted trees I found several diminutive, though botanically interesting, orchid species clinging to branches smothered with lichens but little else of interest. Carefully picking my retreat through the tangle of scrub, I paid close attention to the woodland floor. Coarse grass covered the ground and then I noticed a tiny seedling with maple-like leaves. As no trace of the parent plant could be found, the seedling was collected in the faint hope that it could be kept alive during the next stage of the expedition. Normally such delicate material is avoided unless it can be despatched immediately, and seed is always preferred. But this long shot was to pay off when the plant subsequently proved to be one of the most interesting collections made on this expedition.

A few months later against all the odds the plant flowered for the first time at Kew and was identified as *Abutilon striatum*, a relatively common, tender shrub most often seen in cultivation as the variegated cultivar 'Thompsonii'. However, this form had flowers of a vivid red, with blood-red veining, far removed from the pallid orange of the usual form. In cultivation it is rarely out of flower, an easy and attractive plant for the cool glasshouse or home.

Another plant from this Argentinian expedition which has proved its worth in cultivation is *Glandularia tenuifolia*, a close relative of the familiar hybrid garden verbena. This useful little plant produces tight carpets of finely cut fern-like foliage, above which deep mauve, white-eyed flowers are produced throughout the summer. In the wild the plant has a wide distribution but prefers dry soils disturbed by man. In cultivation it has been found to be an easy and tolerant subject which is a valuable addition to summer bedding displays. It is not hardy and must be maintained by means of cuttings or seed.

The 1985 expedition to Chile led by Stewart Henchie and Tony Kirkham brought back a wealth of interesting species from the vicinity

Seldom seen in British gardens is *Rosularia sempervivum*, a member of the stonecrop family, Crassulaceae. This plant was collected in the Elburz Mountains by J B E Simmons, the curator of Kew, and Hans Fliegner, an assistant curator, during a horticultural expedition to Iran in 1977, and it has been given an Award of Merit by the Royal Horticultural Society.
Photo: J B E Simmons

of Valdivia, Chiloe Island and Lake San Rafael. Amongst them was an exceptionally fine form of *Lobelia tupa*. Lobelia is a popular bedding-plant known to all gardeners; the more discriminating may prefer the various forms *Lobelia cardinalis* and *L. fulgens*, which are perennial plants in various shades of red, but *Lobelia tupa* excels them all.

Few would believe that it is possible to grow 2.5 m (8 ft) lobelias in a British garden, but such is the case with *Lobelia tupa*. The species is just hardy in southern areas but is best given the protection of a wall if possible; elsewhere it will need winter protection. The effort is well worthwhile as the plant is both curious and attractive with its downy leaves and dusky carmine-red flowers borne in terminal spikes at the ends of towering stems. One drawback is that vegetative parts of the plant are reputed to be poisonous; however, many common garden plants possess this flaw. The plant collected by Henchie and Kirkham is a particularly vigorous form. Perhaps it needed to be since it was found growing on a well-drained grass hummock completely surrounded by marshy land, in an area which subsided after the great earthquake and tidal wave that struck Valdivia in May 1960.

Whatever the interest, botanist, horticulturalist, professional or amateur, all lovers of plants and gardens constantly seek the new and unusual. Kew with its great tradition of exploration has for more than two centuries been devoted to meeting this demand in a responsible way. Not for individual gain or pleasure but as part of a greater effort to gain knowledge of the plant kingdom as a whole. Today the tradition continues, but there is now more than ever an urgency to investigate the flora of the world as it is increasingly threatened by a great variety of human activities.

It is already too late for some areas: they have been devastated before they, or the plants which they contained, could be evaluated. Botanic gardens no longer have the luxury of time. In many parts of the world plants collected one year may be extinct the next. If we wish to

Seeds of an exceptionally fine form of *Lobelia tupa* were collected in Chile by a Kew expedition in 1985. It is hardy at Kew and this plant was drawn by Joanna A Langhorne in 1987 and published in the *Kew Magazine*, 1988, Plate 112.

do more than maintain a 'museum' of plants in the artificial surroundings of our gardens then every effort must be made to ensure their survival in the wild. This can only be achieved by determined efforts on a worldwide scale to reduce the pressures which place plant life in jeopardy.

13 *Travel and expeditions by Kew students*

LEO PEMBERTON

From the earliest days of the Royal Botanic Gardens, young trainees have been involved in expeditions and plant collection. For students attending the Kew School of Horticulture, the prime attraction for these young men and women, whose ages range between 18 and 30, has always been the opportunity to work with Kew's extensive and varied collection of plants. This has further stimulated their interest and involvement so that many of them have wished to see plants growing in their natural habitats. There are excellent opportunities for students to participate in expeditions and also to visit and be actively involved with botanic gardens overseas, especially where notable specialist collections are maintained.

When the Kew Horticultural Diploma Course was established in 1963 by Sir George Taylor, there was an official scheme whereby students could apply competitively for a three-week exchange period. Students were free to choose where they spent the time but had to make a reasoned application to a selection panel consisting of staff and students chaired by the Supervisor of Studies. The whole process of applying and selecting gave an extra experience within the framework of the course. Although the original idea was for an exchange between Kew students and students or young staff from other centres, this has not always happened as only a small number of institutions have actually sent students to Kew – the Botanic Gardens at Aarhus, Copenhagen, and the National Trust for Scotland Garden at Inverewe are notable amongst them. On the other hand, there is never a shortage of applications for the six travel scholarships for the Kew students. Many mutually beneficial links have been forged with a large number of West European establishments, botanic gardens and private gardens, and where there has, in fact, been an exchange of staff, this has been an added advantage.

Travel scholarships

Successful applicants for student exchange (now termed travel scholarships) must ensure that they make strong personal contacts with their potential host-centre as well as securing help and advice from senior staff at Kew. These personal contacts greatly heighten the student's awareness of the multiplicity of study and research being undertaken at Kew. This proves beneficial in both directions because even a single visit by a student to a specific botanic garden can materially aid the work of botanists, for example by producing lists of specialist plant groups which are being cultivated at that garden.

On other occasions students have collected herbarium specimens of required genera. In Sri Lanka during July 1986, Sarah Rutherford made a collection of palms, including all the species growing at the Royal

Another Kew diploma holder, Pamela Holt, in Peru during the Machu Picchu Venture in 1981. She is seen holding the orchid *Sobralia*.

Kew horticultural students are often attached to other botanical gardens or notable estates, such as Schloss Schönbrunn, seen here with its rose pergola and ornamental bedding, which was visited by Louise Bustard in 1985.
Photo: L Bustard

Botanic Gardens and Peradeniya as well as at other centres on the island, and Gavin Smith, in the following December, also made a specialist collection of Palmae from Thailand, including rare and little-known species. In these and other instances, the students' experience in making a compulsory collection of British weeds as part of their course-work proved an important preparation for more serious collecting, even of palm trees!

Generally speaking, the students who are given the opportunity of a travel scholarship travel singly or as the only Kew student in a group. This is deliberate policy, designed to encourage the student to gain maximum personal gain from the opportunity and increase self-reliance. In some ways the travel scholarships are an extension of the pre-1938 schemes, whereby students on completion of the two-year Kew Certificate course could spend a post-training year at another notable European garden. Popular destinations were the famous gardens at La Mortola and Villa Taranta in northern Italy, owned by Englishmen who grew a wide range of exotic plants in a very pleasant climate.

Another link with the past is through a student prize named after Frank Kingdon Ward (1885–1958), the famous author and collector of plants, mainly in the southwest of China and the adjacent areas. The annual Kingdon Ward Prize is awarded to the Kew student presenting the best final-year project on horticultural management.

One of the greatest supports for student travel-scholarships is the association of old Kewites – the Kew Guild. Since 1976, through its award scheme, the guild has provided supplementary funding for a wide range of travel opportunities. The scheme was established by generous donations from old Kewites and other benefactors to provide a capital sum from which the interest would be available to aid Guild members with the costs involved in travel and expeditions. Many students join the Guild while they are still engaged on their course of study and the Guild gains from their support. In return, the student members are eligible to apply for awards, and many of them have done so successfully.

The Ernest Thornton-Smith Scholarship

The most prized award for student travel is the Ernest Thornton-Smith Scholarship. This was awarded for the first time in 1968 from funds provided by the businessmen and benefactor Ernest Thornton-Smith, and after his death by the trustees. Initially the scholarship was for a student to study tropical plants *in situ* in tropical South America and the West Indies – the areas where Mr Thornton-Smith had spent his early working-life.

The aim of the scholarship was to encourage a sense of adventure and achievement in the recipent, which would later be reflected in work in the community. During Mr Thornton-Smith's lifetime the student's objectives were not as important as the likelihood of a fascinating account for the patron to read on the completion of the trip. His wish, in part, was to relive his own adventurous past through their experiences and to satisfy the deep interest in tropical plants which he acquired in later life.

After his death the purpose of the scholarship changed and the final written account was no longer the main objective. Busy trustees meeting only once a year did not want to read arresting narratives; instead the applicant's basic aims became critical, and on completion of the journey an accurate statement of expenditure was required! It might seem that the scholarship was primarily concerned with plant collecting but this is not so. Most recipients do bring back some plant material but today this is a very selective range requested by senior staff; haphazard plant collecting is no longer required nor encouraged. The definition of what should be collected (and allowed by local officials) is very precise and students are made aware of the restraints that are imposed by official regulations. Training students to recognise and understand the importance of, and reasons for, government regulations concerning the

Robert Mitchell with a group of Solomon Islanders on Ropongga, and an Australian colleague, during his orchid-hunting expedition in 1986.

despatch of plant materials from other countries to the United Kingdom is essential.

In the early years of the scholarship, students went to the Caribbean Islands and South America, and the tours gradually increased in horticultural content and study. The destinations of the holders of the Ernest Thornton-Smith Scholarship read very much like a travel agent's brochure – Antigua, Borneo, Brazil, Chile, China, Dominica, Ecuador, Guyana, Hawaii, Indonesia, Jamaica, Malaysia, Martinique, New Zealand, Papua New Guinea, Seychelles, Solomon Islands and Western Australia. To date it happens that each locality has been visited only once.

Choosing a destination is an interesting exercise because so many different factors are involved. Senior staff may encourage visits to particular areas because of their personal scientific work, and they will usually have local contacts and sources of information, invaluable when working out an itinerary. Of course the wide-ranging expertise of the establishment is available to students, and the unique library facilities – consisting of detailed maps, books about exploration, local *Floras* and archival material relating to the local institutions – plus correspondence files relating to the knowledgeable and important people in any specified area can be examined.

All the work entailed in making plans for an itinerary with clear and realisable objectives, including possible routes and locations with anticipated centres, must be carefully worked out. Kew's internal sources of information, available from the Herbarium, are just as important as the local ones since the herbarium staff have divided into geographical responsibilities and knowledge which helps the student enormously.

Other funding

Additional travel funds have been provided annually in more recent years with the Hozelock Prize and the Henry Idris Matthews Scholarship, and because this latter award has been made through the Bentham-Moxon Trust, a Kew-based charity, extra funding has been possible from other bodies. Kew students have also benefited from the Alpine Garden Society, which annually awards national competitive scholarships geared to its special interest, and the Studley College Trust, which provides funds for more wide-ranging studies. Numerous other organisations have given financial support when the application has relevance to them. This has meant that students have been able to visit such places as Cameroon, the Canary Islands, Israel, Japan, Mexico, Thailand and Uttar Pradesh in India.

So far the benefits of individual travel have been discussed, but since 1979 there has also been an annual group study-visit for the final-year students to various places in Europe and in Britain. Early encouragement was provided by the German Academic Exchange Scheme (DAAD), which provided a guide–interpreter and very helpful financial support.

Visits were made to the Bonn National Garden Festival and the following year to the Ruhr region. In 1985, a converted double-decker bus was used as a mobile self-catering hotel to take the group to the National Garden Festival in Berlin. This was a very memorable occasion – not just because of the visit to the divided city of Berlin, but due to complexities of feeding a 'multi-race' bus load and crossing frontier borders with many different passport holders. What started off as a group visit rapidly became an expedition! That introduction to European garden festivals undoubtedly contributed to the School of Horticulture's prize-winning entry in the 1984 Liverpool International Garden Festival with their Water Margin Garden.

★ ★ ★

Kew students who have travelled abroad or, on completion of their studies, have taken up posts abroad have played very important ambassadorial roles. In many instances weakened contacts have been strengthened. The international network is always enhanced by encouragement and support, and because the students make these contacts at a low level initially, but with great enthusiasm and ability, many lasting friendships and exchanges of information and plant materials ultimately result.

(*left*) A *Puya* species being admired by Colin Porter during his visit to Peru in 1983 after the diploma course. Offsets from this plant are now in cultivation at Kew.

(*below*) Riverine forest at Dhokal in Somalia, with the fan palm *Hyphaene coriacea* prominent on the skyline, was the site for an investigation of the forest's status for conservation purposes by Mike Maunder, following his course in 1986.
Photo: M Maunder

14 Plant collecting for the Jodrell Laboratory

DAVID F CUTLER

Research work in the Jodrell Laboratory at Kew normally requires the use of live plants. The main areas of work are divided into four sections. The plant anatomy section is where the plants are examined for microscopic details of their structure. Also in this section studies are made on plant breeding systems. The second is the cytogenetics section, in which the structure, function and behaviour of chromosomes are studied with a battery of modern techniques. The third section is that of biochemistry where some of the vast range of chemicals found in plants are extracted and looked at for properties of possible use to man. In addition, there is the work of the Seed Bank in the physiology section, referred to in Chapter 15.

The nodding flowers, with beautifully marked silvery inner tepals, of *Tigridia ehrenbergii* (Iridaceae), collected from Hidalgo, Mexico. The group to which it belongs, Tigridieae, is the subject of an anatomical and cytological survey by Dr Paula Rudall and Dr Anne Kenton at the Jodrell Laboratory.
Photo: P Rudall

Although the living collections at Kew can provide some of the plants needed for such studies, it has often become necessary to obtain large numbers of new plants. The best way to do this is to collect them, and the best people to collect the plants are generally those who will be working on them. They know exactly what they want and can often obtain a great deal of very useful information about the natural growing conditions. This is not only vital for successful cultivation but is also very important in understanding more about the ecology of the plants. And it gives the opportunity to see how common the species is and learn more about its distribution. This may be essential if additional material is needed, particularly if some important discoveries are made about the plants in the research that follows.

In all good scientific research, it must be possible to repeat the observations. It is, therefore, most important that the plants worked on should be of known, wild origin. There is always a danger that plants growing at Kew result from crossing between neighbours, or have been selected for particular properties, or that the records of their origins are incomplete! Such plants are of little value for research of the types we do.

Jodrell Laboratory staff have collected as individuals, as members of Kew expeditions or as leaders of expeditions. In this chapter the emphasis is on collection of plants rather than seeds, although often seed is also collected for immediate growing or analysis, but not in the quantities needed for the Seed Bank. Staff have visited many South American countries and parts of Africa. A selection of examples has been made to show a range of diversity and interest, and some of the fascinating results of subsequent work on the plants are mentioned.

Dr David Hunt with *Tradescantia laxiflora* at its type locality, Cerro San Felipe, Oaxaca, Mexico in July 1976. This species, described a century ago by C B Clarke, another Kew botanist, had not been re-collected there, and had been reported only once from another locality since its discovery 150 years earlier. Studies at Kew on the new material revealed that it differs from the true tradescantias in its chromosomes and several other respects, and it was moved to a new genus, *Gibasoides*, in 1978.
Photo: Margaret Johnson

Collecting in Mexico

In the early 1960s, Professor Keith Jones (former keeper of the Jodrell Laboratory and deputy director of Kew) began the research programme on chromosome evolution in the Commelinaceae (the *Tradescantia* or spiderwort family), especially the New World component of the family, which continues today. He also forged the link with botanists in Argentina, giving a cytology course there in 1972 and then travelling to collect living material for research; that link has been maintained.

Many of the Commelinaceae transport easily and propogate readily. Much of Kew's experimental material was initially provided by the late Dr Hal Moore of the Bailey Hortorium, in the United States, who had collected principally in Mexico. Identifications of other material already in the Kew collections were provided by David Hunt of the Herbarium staff, but uncertainties over the wild origin of the old material soon made its replacement essential. David went to Belize and Mexico on four occasions with this primary objective. Cytological studies made on these plants, and subsequent specimens collected in Mexico, have proved most exciting, giving clues to the ways in which the chromosomes have evolved in these plants as well as helping to answer many questions on their classification.

Further material was soon needed and the first of a series of Kew expeditions to Mexico was organised in 1976. Keith Jones, Simon Owens and Margaret Johnson (who went at her own expense) took part, with David Hunt as guide and taxonomist. They were followed in 1979 by Simon Owens and Ann Kenton from the laboratory, Peter Gibbon from the Living Collections Department and by Tony Howard, a friend and former Kew horticulture student, travelling privately. It was on the second Mexico expedition that Tony had a nasty surprise! He was minding his own business when a large lizard chose to run up his trouser leg. He had to call for help, because he needed both hands to keep it from going all the way up!

In addition to the usual collecting records, Peter Gibbon took detailed notes on the growing conditions of the specimens for subsequent cultivation at Kew. He was equipped with an altimeter, a whirling hygrometer for measuring atmospheric moisture and a soil-testing kit, complete with conductivity meter.

Simon Owens collected plants for his studies on breeding systems. His research involves work with the transmission electron microscope. He needs live plants which are specially prepared for examination at

Map 14 Chile and Argentina showing the place-names mentioned in chs 12 and 14.

very high magnifications. More recently, Margaret Johnson has extended her bulb collecting to Turkey, with Professor Neriman Özhatay of the University of Istanbul. She has drawn some interesting taxonomic conclusions from her cytological observations.

A third visit to Mexico was made in 1983, by Ann Kenton and Paula Rudall from Kew, with Tony Howard. They joined forces with Guadalupe Palomino, a cytogeneticist who knows the region well. They were interested in collecting *Sisyrinchium* and species from the Tigridieae (Iridaceae). The attractive Tigridieae in Mexico are particularly interesting because there are numerous microspecies or local variants that do not hybridise with one another.

Collecting bulbs and rhizomes in Argentina

I maintained the Argentinian link, when in 1976, after giving an advanced course in plant anatomy to botanists in the Instituto Darwinion, San Isidro, Buenos Aires, I went into the field with Professor

Mujica to collect bulbs and rhizomes of Liliaceae, Iridaceae and Amaryllidaceae. First the journey took me south, ending at Bahia Blanca. I was surprised to find cacti growing within the high-tide zone of the beach there. But, of course, cacti can withstand drought, and a number grow in saline soils, from which it is difficult for plants to extract any water that may be there.

Most of the area Professor Mujica and I had travelled through in the Citroen 2CV was very flat; the head-winds were often strong enough to restrict our top speed to about 80 kph (50 mph). There is one area of hills though, in a region which is otherwise famous for its plains and wide skyscapes. This is called the Sierra de la Ventana: in fact, there is a window-like opening in the rocks at the top of one of the hills. The flora on the hill slopes was rich in petunias and lobelias, ancestral species of many of our garden varieties and forms.

Bulbous plants are easy to collect in the soft and sometimes moist soils of the south. The job becomes much more difficult in harder, sun-baked soils further north. In these circumstances, a stout screw-driver is more efficient than a trowel. We had to be very careful not to damage the bulbs, and a fairly wide hole had to be excavated, bit by bit. Sometimes we had to resort to the use of a trenching tool when the soil was set like concrete.

For example, it took about an hour to chip the soil from around the large, succulent root tubers of a rare *Herreria* species (*H. montevidensis*). The effort was worthwhile, though; the plant is growing well at Kew and has produced fertile seed. We have been able to raise seedlings for distribution to other European botanic gardens. This is a strange plant, related to members of the Liliaceae. It has a basal rosette of leaves when young, but soon long, climbing stems are developed, similar to those of some *Asparagus* species. These bear whorls of smaller leaves on short lateral shoots and insignificant pale yellow flowers.

The return journey to Buenos Aires was made by air. After a few days of preparation, I set out again, this time to the north, and with three botanists: Rosa Guaglianone, Alicia Rotman and Isobel Casabona. We had a pick-up truck, with room for three in the cab, and one unfortunate in the hard-covered back, among the tents, food-preparation gear, supplies, plant presses, drying papers and drying stoves, not to speak of the luggage. At that time there were frequent security checks on the road, by the police or military. They could not make out what we were doing! On one occasion I was asked if I was travelling with my family, and to try and explain why we had a vanload of dried plants was not easy.

The route took us through Entre Rios Province, with the Parana River to the west and the Uruguay River to the east. In recent geological times, the area chosen for the main road in the east of the province appears to have been a riverbed. It is composed of small- to medium-sized pebbles, rounded, loose and easily thrown up by wheels, particularly those of large lorries. Known as the *ripario*, these pebbles would smash the windscreen if a stout wire-mesh protector were not fixed in front of it.

(*left*) This *Cypella* species, in the iris family, was collected from an open area in Misiones Province, Argentina. Although the soil was shallow, the bulbs were very deep, between the fissures in the rocks and they were dug up with the careful use of a large screwdriver.
Photo: D F Cutler

The striking flowers of *Sprekelia formosissima* (Amaryllidaceae) can be seen from a distance in their natural habitat. The plants are 60–80 cm (24–31 in) tall, and the bulbs up to 20 cm (8 in) deep in the wet ground. They grow in cultivation, but find it difficult to flower; it is very hard to simulate their natural environment. This specimen was collected from San José, Misiones Province, Argentina.
Photo: D F Cutler

The next province north is Corrientes. To the north it borders on the Alto Parana River, which forms a natural boundary between Argentina and Paraguay. The province consists essentially of a series of vast slow-flowing shallow lakes, the *esteros*. Most roads are built on raised banks, to remain above the seasonal floods. The area is populated by large herds of cattle that seek pasture where they can. There is a wealth of marsh and aquatic plants, and a very extensive bird life. The oven birds deserve special mention. Their nests are built of mud, and are common on the tops of fence posts and telegraph poles as well as in trees; the small entrance hole faces away from the direction the sun reaches at the hottest time of the day.

An isolated outcrop of rock in the middle of the swamps was our target. The three low hills, called appropriately Tres Cerros, are of particular botanical interest because they are widely separated from other hills of similar rock, and have a completely different type of vegetation from that of their wet surroundings. They have, for example, a number of cactus species, and I was able to collect several members of the iris family.

To the north and east of Corrientes is Misiones Province. This has a more tropical feel about it; the soils are red laterite for the most part. It is extensively forested; parts had been cleared for the ubiquitous cattle farming, but now large areas of the rich natural forest, with many broadleaf trees and the monkey-puzzle relatives (*Araucaria*) and podocarps (*Podocarpus*) have sadly been felled and planted with the pine *Pinus caribaea*. Needless to say, the rich epiphytic flora of the broadleaf trees is rapidly becoming extinct.

Fortunately the gallery-forest areas around the famous and spectacular Iguazu Falls are in a national park, but the safe area is all too small. The forest did not provide many of the species I was looking for. However, there were some good specimens of *Trimezia martii*, an iris-like plant, and offsets of these have not only proved to be of scientific interest, but flower for the pleasure of visitors to Kew in the Temperate House.

(*opposite page*) The road between Cachi and Salta, Argentina, winds through the mountains on the eastern side of the Andes. The rocks are very soft and subject to erosion and landslides, which often block the road. The large columnar cactus, usually known as *Trichocereus pasacana*, dominates the landscape – it is in danger of extinction since it is widely used for timber in local buildings.
Photo: D F Cutler

In Misiones Province, Argentina, near San Pedro, the natural, seasonally wet forest is being felled to make way for pine plantations to feed paper mills. Expedition members were able to collect many live epiphytic plants from the fallen trees, and they are now growing at Kew. Epiphytic species depend on particular sorts of tree for their survival in the wild, and are liable to become extinct when their natural habitats are destroyed.
Photo: D F Cutler

Collecting in the Andes

On the second occasion I visited Argentina, in 1978, it was as leader of a collecting expedition with a group of Kew colleagues, with a much wider collecting brief. As previously, we had first-rate collaboration from the Argentinian botanists. Duplicates of all the herbarium specimens collected were deposited in the herbarium of the Darwinion. The director, Dr A Cabrera, very kindly allowed us to use one of his institute's vehicles, a hardy 2-ton Chevrolet pick-up truck with a hard cover. He also helped by permitting a series of his staff to accompany us for two-week periods in the field and lending the collecting and drying equipment. This latter saved a great deal of time and expense and certainly avoided the problems with temporary importation of equipment that usually plague expeditions.

The Kew group included horticulturist John Lonsdale, whose main aims were to collect live plants from the Cactaceae, and epiphytes, particularly from the Orchidaceae, Bromeliaceae and Cactaceae. He and I tended to work together, since I was again increasing our collection of live research material with bulbous plants. The other members were from the Herbarium; in addition to making general collections, Steve Renvoize was concentrating on grasses and Melanie Wilmot-Dear on Compositae. We collected in Entre Rios, Corrientes and Misiones and then drove across the deep dust roads of the Chaco and collected extensively in the areas round Tucuman, Cafayate, Salta, Jujuy and Ledesma.

On the journey the road climbed steadily from the lowland plains into the Andes. It became precipitous in places, and was in need of constant repair. There was one stretch where we had to drive on an improvised road through the flood plain and bed of a mountain river since there had been extensive landslips and the entire road was being rebuilt. We had to keep crossing and recrossing the winding river, driving axle deep through the water on large, rounded stones. To make life more exciting, the dynamo had failed, and we were driving on the battery! Eventually we had to climb out of the course of the river up on to the part-made road, to go over a mountain pass. The road was blocked by two working bulldozers, and as we waited for them to smooth a way the engine gave up. It was as well that the vehicle was robust because we needed a long push-start from one of the bulldozers to get us on our way.

The mountain scenery is wild and beautiful; frequent landslips expose red, orange, yellow and occasionally slopes green with copper salts. After collecting in the area round Tucuman and Jujuy, we made our way to the town of Humahuaca, at an altitude of about 3000 m (10,000 ft) and then climbed a thousand metres higher a little further north. The air is particularly poor in oxygen in this part of the Andes, and visitors frequently suffer from what is known as the *puna*, a form of altitude sickness – *puna* also being the term for the type of vegetation found there. Our guide made us stop frequently and leave the vehicle to walk as we gained altitude, to help us get used to the conditions, and we saw the remains of Inca dwellings and monuments.

We also collected numerous plants; it is a remarkable area for Cactaceae and bulbs. The days are very hot, and the nights extremely cold. There are saline lakes, the *saltinas*, and a number of salt-tolerating plants of interest were collected there. Great contrasts are found in the vegetation, depending on the aspect of the mountain slopes. West-facing slopes were all arid in the rain shadow. To the east, near Ledesma and particularly on the east-facing slopes, it was considerably wetter, and the mountainsides were forest covered. John Lonsdale was in his element, with cacti on the one hand, and epiphytes on the other.

Other South American expeditions

Ann Kenton from the cytogenetics section collected in Peru with Paula Rudall from the plant anatomy section in 1985; they continued their joint studies on the breeding systems of plants that started with the Mexican collections, the one looking at their cytology, and the other their embryology and vegetative anatomy. They obtained good material of the Commelinaceae and Iridaceae, from a range of altitudes. On this occasion, they were also collecting reference material of useful plants for the Kew museums. These were bought from the market stalls, and included the rhizomes of *Ullucus tuberosus*, that swell like potatoes, and *tarwi*, (*Lupinus mutabilis*) seeds that are boiled, then washed for three days in the river to remove poisonous alkaloids.

In 1988 Paula Rudall also collected in Brazil, in an area I had visited the year before. It seemed to me to be very rich in Iridaceae; this proved to be true, and she and Brazilian colleagues found several species new to science. Collaboration with botanists at São Paulo is very good, and Kew enjoys a special relationship with the botanical institute of the university there. Dra Nanuza L de Menezes, in particular, works closely with staff at the Jodrell Laboratory. She is writing part of the account of Amaryllidaceae for a volume in the series 'Anatomy of the Monocotyledons' for us.

Jodrell expeditions to Africa

The first Jodrell expedition to Africa involved Peter Brandham and myself. We had been working on the Kew collections of *Aloë, Gasteria* and *Haworthia* for a number of years. Peter studied the chromosomes and I looked at the fine structure of the leaf surfaces as seen with the scanning electron microscope. The adaptations in the leaf-surface structures were clearly related to fitting the plants for survival in a range of different habitats. In our joint research, we were aiming to see if the intricate surface features were strongly inherited or if they could be altered by the environment. It was very important that we should see the exact growing conditions of the plants, and we needed many more samples to be sure that our observations were sound. Peter was also particularly interested in the part played by the increase in chromosome numbers in the evolution of the aloes and haworthias.

The dominant species of aloe on the hills near to Uitenhaige, South Africa, is *Aloe ferox*, which is a very widespread species. It was collected from over a large part of its range, to look for possible variation in its chromosomes and microscopic features of the leaf surface in a major survey by Dr P E Brandham and Dr D F Cutler.
Photo: D F Cutler

Our aim was to collect very widespread species from across their natural range so that we could look for variation or stability in their characters. We also wanted to collect species that have very limited distribution, and very specific habitat requirements. We started in Kenya. Here we collected aloes along and up the rift valley. We visited habitats near Kisumu and on the slopes of Mt Elgon, and our route took us to Isiolo and Nyere, a few miles from Mt Kenya.

This was after we had bailed out the Kew-owned Toyota from a garage in Nairobi! Some essential work had been done on the suspension, but payment had not come through from London. It was only because a botanical colleague at the Nairobi Herbarium lent us the money to pay the garage that we were able to make the journey at all. The vehicle gave us some memorable moments, mainly because it was not very good at stopping.

Although we carried two spare wheels, we were left one Saturday evening with three flat tyres. This was partly due to the sharp volcanic stones on roads near Archers Post, and partly to little boys said to be commissioned by a garage proprietor to put nails on the road. The puncture repairs failed after about 80 km of pounding. Dusk was falling on a remote road, when a car stopped for us. It was crammed full of Kikuyu in their best clothes, ready for a Saturday night out. Somehow they accommodated me plus wheel, and took me about 16 km (10 miles) to the nearest garage. They were incredibly kind being prepared to take me back to the vehicle if the garage mechanic could not.

After sending the extensive collections back to England by air-freight, together with their plant health-certificate, we ourselves flew on to South Africa. Hiring a VW Combi in Pretoria, we set off on the long southern road to Cape Town. We had arranged to meet a series of *Aloe* and *Haworthia* enthusiasts *en route*, and they were all most helpful.

Some of the localities are very difficult to find, and are often on private land, so it was essential to have planned the route well ahead and our contacts knew the landowners. We were well received everywhere and enjoyed excellent hospitality. The person who accompanied us for most of the journey was Bruce Bayer, the *Haworthia* expert, and then Curator of the Karoo Garden at Worcester, Cape Province. He has an extensive knowledge of the area and the best localities. Many of the *Haworthia* species we were looking for were very small and some have only the leaf tips above the ground. They are very difficult to find, even if you know where they grow. We made extensive notes on each plant we collected, and took photographs of the plants themselves in close-up, and of the habitats.

The research on these specimens has proved to be very interesting and has produced a series of papers. It is evident that the leaf-surface features are under strong genetic control, are not influenced by local variations in environment and are vital to the survival of these plants in very hot, dry and hostile conditions.

I have visited South Africa twice since our first collecting trip. Peter has been to Kenya, Tanzania, Ethiopia and Somalia since. Margaret Johnson from his section and Susan Holmes from the Herbarium accompanied him on some of these expeditions. On one occasion in Somalia, Peter was offered some camels in exchange for Margaret!

Kew now holds one of the most comprehensive collections of the Aloaceae in the world as a result of our extensive collecting. The plants are of particular scientific interest because their wild origins are known; most of those in other collections outside Africa are poorly documented. The research has extended into biochemistry, where Tom Reynolds and his team have isolated a number of interesting and potentially useful chemicals. He has also shown that the same chemicals

Aloe, *Gasteria* and *Haworthia* species collected from their natural habitats are grown for research at Kew in special greenhouses not open to the public. The attractive orange flowers of the aloes can be seen here. Also, the small plants in the foreground were raised by crossing different *Gasteria* species for research on their breeding systems and chromosomes.

can be used to help our understanding of the inter-relationships of the aloes. Susan Holmes's main interest is in the Euphorbiaceae, and she was able to collect a good number on the expeditions because they grow in very similar conditions to those where aloes are found. She has collaborated in a number of papers with Jodrell staff of all disciplines – a good example of integrated science at Kew.

Growing plants for the laboratory

Live plants that result from the various collecting trips are grown at Kew by specialists in the greenhouses devoted to scientific collections. The collectors' notes and personal comments ensure that the most appropriate conditions can be provided for good growth. There are always moments of delight when rare species are brought into flower. When their scientific purpose has been served, the plants are either retained as specially important collections, or they may be grown on in the public glasshouses if they are worth displaying. Herbarium voucher-specimens are made as a permanent record for posterity, if they were not made at the time of collection. Species that have been shown to be rare and which have multiplied are distributed for conservation to other botanic gardens.

15 *Kew Seed Bank collecting*

SIMON LININGTON

Since Kew was founded in the 18th century, seeds have been the main means of transporting new and exotic species for growing in the gardens. Easier to ship than cuttings or other vegetative material, seeds also provide the means by which whole populations can be genetically represented and stored alive for tens, if not hundreds, of years. Samples can be drawn off from such conserved material for research into a myriad of subjects or for direct incorporation into breeding and agronomy programmes beyond Kew. Such is the basis of the Kew Seed Bank, located at the satellite garden of Wakehurst Place in West Sussex.

Kew Seed Bank containers. A whole plant population can be genetically represented within a very small volume by means of seeds. Storage for decades if not centuries seems feasible for many species using seeds carefully dried and held at sub-zero temperatures. Such is the scientific basis of seed banks which act as a biological safety net against the loss of species from the wild. *Photo: Kew Photo Unit*

Kew's Seed Bank evolved from the need to deal more effectively with the annual deluge of requests for material via the international exchange of seed lists (*Indices Seminum*) between botanical institutes. Work involved in the annual collection of seeds from the herbaceous beds and individual plants growing in the Gardens was reduced by collecting a large amount of seed and holding it under tolerably good storage conditions for several years. This change coincided with the improved understanding of seed storage by researchers, such as those at Reading University. By careful reduction of the moisture content under low temperature and low humidity, followed by the freezing of the dried seeds, long-term storage appears to be practicable for a wide range of species. With these improved techniques came an increase in research within Kew's Jodrell Laboratory into the important related subjects of seed dormancy and the storage of the so-called 'recalcitrant' species, to which drying or freezing spells death, such as oak acorns (*Quercus*).

Consequently, the Kew Seed Bank was well placed when the main surge of interest in the *ex-situ* conservation of wild species occurred in the early 1980s. Crop gene-banks had been in existence for years, promoted by a network set up by the International Board for Plant Genetic Resources (IBPGR) based in Rome. The concept of a bank solely devoted to wild species was enhanced by the work of Dr Peter

Thompson, who was in charge of the Kew Seed Bank in the early 1970s. From its inception the Bank turned its attention away from botanical garden-collected seed, where the risks of hybridisation between plants of different provenance growing in close proximity threatened the integrity of the collections. Similarly, collections from single plants were often all that was possible, thereby limiting the genetic usefulness of material distributed.

The trend was towards collection of seeds direct from the wild and, when multiplication was necessary, to carry this out under isolation conditions. But what was to be collected? With so few workers internationally involved in the collection of seeds for long-term storage, there was initially the target of the entire world's flora, nearly 250,000 species, many of which were under threat and few of which had been made available to research!

A wide range of species have potential as forages in the tropics. In collaboration with the Brazilian genetic resources organisation, Kew collected seed and vouchers of diverse wild species from the northeastern state of Bahia in 1983.

The current objectives of the Kew Seed Bank

The loss of species from the tropical rainforests is well documented, but the problems of seed conservation even within a small area are immense. A large number of the primary rainforest species would appear to have large fleshy seeds and this is often a strong indicator that they are recalcitrant. Long-term storage of such species is thus impossible with present technology. Added to this, collecting work is difficult due to the great variation in flowering and hence seeding times for a given species, which would seem to necessitate the placing of workers for long periods in the field, and also due to the tendency for populations to exist as single trees scattered over large areas of rainforest. These problems are being considered but *in situ* conservation is the only real solution in such areas at present, especially when one considers the difficulty of growing many of the species in ecological

Seed collecting in Zaire. A joint collecting team consisting of four nationalities gathering forage germplasm in Park Virunga, in an area regularly grazed by hippopotamus herds. Apart from seed returned to the Kew Bank, material from this 1983 expedition has been incorporated into breeding programmes within Africa.
Photo: J Terry

isolation. However, this scientific exploitation is likely only when material is also able to be stored *ex situ*.

A further obvious area for work is the dry tropics, where important populations, if not whole species, are threatened by a combination of human pressure and climate interwoven to produce desertification. By saving representative collections from these areas, they can be made available to research and their potential determined either as forages, land stabilisers, hedges, sources of disease resistance for related crops or as medicinals, especially if local plant use points in that direction. More formal plantings of promising species either at a village level or on a larger scale could have an important bearing on these hard-hit areas.

The international agricultural research centres, such as the International Livestock Centre for Africa (ILCA), are leading the way in this work, especially with forage and browse species. However, the problem is so vast and the range of species as yet looked at so limited that the Bank, especially with its links to herbarium taxonomic expertise within families such as the *Leguminosae* and *Gramineae*, has a valuable role to play at the conservation level.

Following collecting trips to the Mediterranean basin (mainly Greece and Italy) during the 1970s, when a large amount of valuable material was collected, attention has been turned more and more to the dry tropics, where seed collecting is effective due to the usually clearly defined seeding period and a good response to the drying and freezing methods employed by the Bank. A preliminary expedition to Kenya in 1981 gave valuable insight into aspects of large-scale seed collecting in the tropics and some useful measurements of seed-viability loss under different conditions were obtained. This was followed by a joint Kew/Brazil expedition within the northeastern states of Bahia and Piaui, when the emphasis was on potentially useful dry-land forage genera such as *Stylosanthes*.

Trees of *Conocarpus lancifolius* growing beside a dried-up river bed in Northern Somalia where it is probably endemic. Apart from being an excellent shade species in this dry region it can be used for producing poles, and also has value medicinally and as a fodder. Seed was collected from this stand in 1983.
Photo: S Linington

The same year (1983) there was an expedition in conjuction with the National Range Agency, Mogadishu, to Northern Somalia in the Horn of Africa to look for a range of species identified by the arid-plants data base (now called the Survey of Economic Plants for Arid and Semi-Arid Lands, SEPASAL, based at Kew). Seed of a number of interesting species was collected, including that of the trees *Conocarpus lancifolius* and *Mimusops angel*. Somalia is noted for its range of endemics. Further African trips have been made possible by joining IBPGR-organised collecting trips to Zaire, Burundi and Rwanda (not admittedly arid areas) and to Mali, where collections were made in the Sahelian region.

Owing to the limited number of Bank staff, the fact that collections are continually arriving as a result of expeditions external to Kew and the time necessary to maintain existing collections, opportunities for collecting by staff are limited. Planning an expedition and the associated work after its completion can sometimes occupy nearly as much time as the trip itself.

Planning a seed-collecting expedition

Before a trip has been approved within Kew on the basis that it fulfils the Bank's collecting policy and that there are likely to be good returns on a large investment of time, manpower and money, an approach will have been made to the appropriate authorities within the country to be visited to ensure the visit would be welcome. Seed collecting has a range of political implications, as has increasingly the collection of all plant material. Plants are now viewed widely in the same light as any other natural resource. By ensuring free exchange of material and information between countries and also by leaving a representative sample of seeds or duplicate pressed specimens (voucher specimens) in the country of origin, organisations such as Kew are helping to promote the concept of an international heritage of plant material.

Once all necessary permission has been granted, the usual preparations proceed, such as checking that visas and passports are satisfactory, that all the contacts have been made with participating organisations, that transport is arranged, flights have been booked, inoculations are up to date and equipment has been airfreighted out in advance.

The collectors need to familiarise themselves with the likely target species that they are going to encounter. Although seed collecting can yield useful material by collectors with little taxonomic knowledge other than an eye for distinguishing one species from closely similar ones (thereby avoiding mixing collections), obviously such an approach can lead to a high capture-rate for worldwide tropical weeds. That is not to say that pantropical weeds are not of potential interest, merely that they are easily obtained internationally and the Bank is less likely to fulfil its role of leading researchers into the use of new species.

With some 5 million voucher-specimens, the Kew Herbarium provides an excellent first-stop in the preparation of a seed-collecting trip. The pressed specimens provide not only information on likely seeding dates and the morphology of target genera and species, they also give data on possible locations for collecting. However, it is important to look at the date on the specimens because the world is a much changed place since even the 1950s and the collector might discover that a likely location is now taken over by a sorghum field or a building development. As information systems improve, likely areas to collect will be increasingly defined by reference to a mixture of climatic data, satellite imagery and digital maps, be they of vegetation, land relief or demography.

One aspect of planning a seed-collecting expedition is to consult climatic data in order to predict whether drying agents will be required to minimise losses in seed viability in the field. Relative humidity and temperature together with the seed's oil content affect its moisture content, and the latter is a major component in determining longevity. In the Sahel semi-desert for instance, daytime relative humidities and temperatures are such that 'safe' moisture contents will be achieved. At night, as temperature drops quickly, humidity rises and it is useful to place the seeds collected into cloth or paper bags within a thick plastic sack for these hours. Because the temperatures experienced within an expedition vehicle can vary considerably, the coolest spot (normally low down away from the window) should be chosen for the seeds.

In the wet tropics use of a drying agent is advisable, especially if the chances of freighting material back to the bank are few and far between. Drying agents vary from silica gel to dried rice. Sundrying of seeds should be avoided because, although it promotes a quick reduction of moisture content, the temperatures encountered are also liable to reduce the longevity of the seed intended for storage over 200 years.

Since the seed collections can be multiplied only with difficulty once back in temperate conditions, except in expensive glasshouse environments, it is beneficial to collect enough seed to allow for distribution to the Bank's international users over the life of the collection. Under no conditions is more than a fifth of the available seed collected from a population otherwise the reason for its decline could be seed conservationists! Of course, with some important collections only a few seeds will be collectable, and in this situation multiplication *ex situ* will be necessary.

A further consideration when collecting seed from wild populations in the field is that many seeds are aborted in a bad season and thus the collector should check whether the seeds or fruits being harvested are completely filled. Because the users of the Bank's material could waste months of research time if collections were wrongly identified, a pressed specimen collected at the same time as the seed is harvested is valuable. Of course, specimens at seed maturity are less useful to taxonomists than at flowering and so it is sometimes necessary to grow a specimen for identification on return.

With all such biological collecting, recording data about where and when the material was collected is vital. The location needs to be pinpointed as precisely as possible to allow for a return visit to the site

Tree seed collecting for research in the Solomon Islands. Sometimes it is necessary to bring seeds down to earth (one benefit of following timber-felling gangs). The material was used by Kew physiologists working in the Solomon Islands during 1978 to examine tropical tree seed storage behaviour.
Photo: R Smith

at a later date. This is often quite difficult in desert locations, where major fixed landmarks are few and far between. Local guides, an essential component of such trips, often are able to give local names to the smallest settlement on a dried-up river valley. The problem, of course, lies in the spelling and subsequent location on even the very best large-scale maps. Consequently, many sites are recorded as being a set number of kilometres from village A towards village B. Perhaps technology borrowed from the military for pinpointing land locations will eventually overcome this problem. Other data, such as ecological descriptions and altitude, are useful as they help the user of the material to predict its likely growing requirements.

A seed-collecting trip in Mali

There is no such thing as an average seed-collecting trip for no two are the same. Most in semi-arid countries are more a test of stamina than anything else. Many hours can be spent being thrown around in the back of a crowded Land-Rover on often abominable or nonexistent roads. Being low in humidity, the hot air is not so tiring as in the wet tropics but nevertheless the heat can cause fatigue. If conversation in the vehicle is lively, then the hours pass quickly, otherwise it is easy to find oneself drifting into a reverie relieved occasionally by sightings of interesting wildlife or of plant species in a promising location.

If seeds are available, then the collector in us all takes over and the urge to gather in becomes paramount; the thrill of several well-filled seed bags of a useful species makes up for any of the other inconveniences. Once on the trail, which may take you some distance away from the vehicle, other target species in seed come to light and you are often

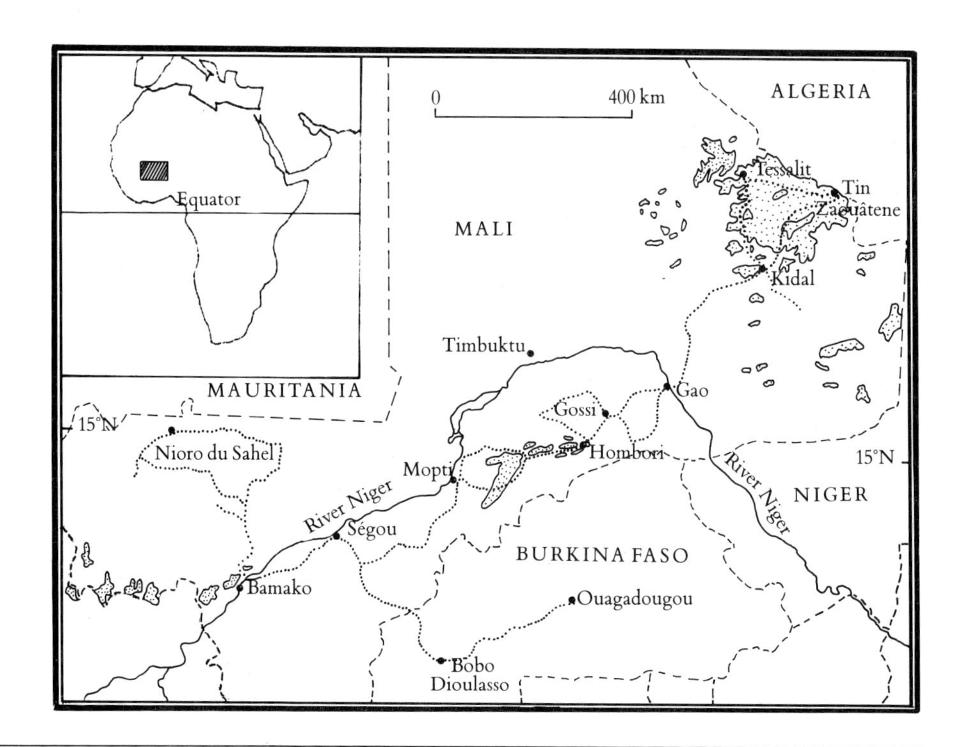

Map 15 Part of West Africa with the place-names mentioned in ch. 15.

The Bandiagara Escarpment in Mali, home of the Dogon people who farm the cliff tops. On expeditions, the sturdiness of the vehicle is all-important. With water and fuel stored in cans on the roofrack, the journey down the rough track from the cliff top was certainly testing.
Photo: S Linington

frustrated by the lack of extra pairs of hands. Even in the most isolated arid location a passing nomad will appear as if from nowhere, watch for a while, even join in the activities and then move on, perhaps convinced that modern travel has opened up his domain to lunatics.

The flavour of such an expedition in the semi-arid tropics might be given by a brief account of aspects of a visit to Mali in West Africa in October–December 1986. The trip had been organised by IBPGR as part of its eco-geographical collecting activities in the Sahel zone. One of the primary aims was to survey and collect populations of the wild and weedy *Pennisetum* grasses. Kew participated in this expedition both to assist in this activity and to collect other species of potential economic interest.

The trip was led by a Belgian, Michel Horn, the IBPGR regional coordinator for West Africa, while the Mali counterpart from the Institut Economique Rurale was a millet breeder, Aboubacar Touré. Like Kew, the International Institute for Tropical Agriculture (IITA), based in Nigeria, had placed a scientist (Abimbola Osunikawa) on the expedition. Together with Bernard, our driver from Burkina Faso, who was particularly good at puncture repairs (we had plenty!), the collecting team comprised five different nationalities.

In addition, a third of the way through the trip, while in the Gourma region, we obtained the invaluable services of Mohammed Ag Nutt Nutt, a veterinary officer based at the nomad encampment of Gossi. His knowledge of local plants and their usage was excellent. Similarly his sense of direction across country with few features other than slight undulation was amazing. Because local languages are used in the wilder parts of Mali we visited, conversations with herdsmen, which always started with elaborate greetings, often reached me via two or more translations.

Later in the trip, as we headed towards the edge of the Sahara we were accompanied by a military guide from the town of Kidal (dominated by its open prison). His local knowledge was excellent and his ability to spot a distant gazelle was legendary amongst the group. Carrying a rather old rifle, he regularly stopped the vehicle to take pot-shots at the wildlife, luckily with limited success. One sandgrouse was, however, reduced to a pile of feathers. Six people in one vehicle for a prolonged time put a strain on the rugged Land-Rover, which was already loaded down with collecting equipment, luggage, fuel cans and drinking water. Some problems were experienced with the vehicle's roof after 12 villagers requested a lift back to their village after helping to dig the vehicle out of appalling muddy ruts on a track we took at the start of the trip.

We set out on our expedition from Ouagadougou, the capital of Burkina Faso. After driving to Bamako, the Mali capital lying on the River Niger, we headed into northwest Mali along roads and tracks badly damaged by rain and eventually along the drought-ridden Sahelian border with Mauritania. Returning to Bamako for more provisions and a brief rest, we headed up the metalled road towards the old town of Gao in its important position on the south side of the Sahara and beside the River Niger. We were accompanied as far as Mopti by a second expedition, which was going to explore the inland Niger delta for wild rice.

From Mopti we headed towards the great escarpment that runs up through that area of Mali. The top of the escarpment is the centre for the Dogon people who farm small fields with soil brought from the plain below and with water from reservoirs created by small dams. After a hair-raising drive along a track down the escarpment and a brief time disorientated on the plain, we reached the table-top mountains near Hombari. From there we spent time exploring the dry Gourma region around Gossi, getting lost in a sandstorm on the night of 5 November. Having slept on campbeds in the lee of the Land-Rover, we awoke to a bonfire lit by Mohammed our guide.

After a brief stop in Gao we headed up to the increasingly dry Adrar des Iforas region of mountains in northeast Mali. Arriving after dark in the village of Tin Zaouâtene on the border with Algeria, we received a particularly cool reception from the frontier guards based in a Beau Geste fort there. This was in contrast to the great hospitality we received at most villages despite sometimes great poverty. Heading westwards we travelled back to Gao along the Vallée du Tilemsi by way of the southern end of one of the main Trans-Saharan routes. On reaching Bamako again, we had completed nearly 10,000 km (6200 miles), well over double the equivalent of the journeys Bamako–London.

In the course of the trip a large amount of seed had been collected. Important wild and weedy *Pennisetum* material had been collected, both for long-term conservation at Kew and other gene-banks, and also for direct incorporation into the Mali millet-breeding programmes. Wild millet is thought to possess good genetic disease-resistance and in some cases is a source of male sterility, important in hybridisation programmes. In addition, some important famine food species had been collected, such as the sweet-fruited *Grewia* species, *Ziziphus mauritania* and the grass *Panicum laetum*, the seeds of which were swept up from the ground where large stands existed. IITA benefited by large numbers of *Vigna* collections made particularly in northwest Mali and collections of wild rice among them a number of pure stands of *Oryza longistaminata* in the Gourma region.

Medicinal plants included the relatively widespread *Guiera senegalensis*, extracts of which are taken as a local remedy for asthma and coughs. Also collected were fruits of the shrub *Cadaba glandulosa*, perhaps the most important medicinal plant in the Gourma region since it yields a treatment for hepatitis; the leaves are dried in the shade, pulverised and then eaten.

★ ★ ★

Apart from an introduction to the Sahel, its people and flora, the trip holds a number of vivid memories for me. For instance a night spent in the chief's hut in a small village a few days after the start of the expedition. As we assembled our campbeds and put up mosquito nets by the light of a hurricane lamp, the whole village watched from the two doorways with much hushed chattering. Then there were the crocodiles kept as sacred animals in the village pond at Diangounte-Camora, the bateleur eagle soaring high over a collecting site, the hundreds of migrating garganey, ruff, wood sandpipers and pratincoles at Gossi reservoir, the large python sliding across in front of the vehicle in savanna scrub, the total absence of traffic on miles of rutted tracks in the northwest and the amazing clear night skies. All unique experiences which added to the excitement of collecting seeds for the future.

16 *Fern hunting in the tropics*

BARBARA PARRIS

The greenhouses at Kew include magnificent displays of tropical and temperate ferns, while the Herbarium houses a research collection numbering tens of thousands of tropical ferns. Even so, new material is constantly sought by Kew expeditions in order to further research and add to our knowledge, especially of the tropical species.

Many botanists who collect plants have been brought up in the north temperate regions and have a rather limited idea of what ferns look like, based on the common northern genera *Dryopteris, Polystichum, Polypodium, Thelypteris* and *Athyrium*, possibly also including *Botrychium, Ophioglossum* and the filmy fern genera *Hymenophyllum* and *Trichomanes*. Thus the general botanical collector who ventures to fern-rich regions will gather the plants which are obviously ferns, but those with fronds not obviously 'fern-like' in shape are likely to be missed: some may resemble orchids or other flowering plants or even mosses. In fact these important members of the tropical rainforests and warm temperate forests are found particularly in montane regions and range in size from diminutive filmy ferns, with fronds less than 1 cm long appressed to bark or rocks, to huge tree ferns more than 20 m (60 ft) tall, such as *Marattia* with fronds 10 m (35 ft) long. Their fronds can be simple or divided as many as five times.

Even when the 'fern-like' ferns are collected, the non-specialist is likely to collect only one or two members of a group of species because to an untrained eye they may look very similar. The non-specialist, in addition to missing species because of unfamiliarity with the nature of fern diversity, will possibly collect sterile material, which is very difficult, if not impossible, to identify, or poor specimens.

With small ferns such as the filmy ferns, Hymenophyllaceae, which grow in mats on trees or rocks, it is very easy to make mixed collections of several species. These mats of material need careful disentanglement before pressing. Few things are more annoying than

Oleandra pistillaris produces whorls of simple fronds on stems which are upright or pendulous in mature plants. Young plants have creeping stems which produce roots at close intervals and are firmly attached to their substrate which is usually a tree trunk or a rocky bank. The species occurs in Java and Peninsular Malaysia: the plant illustrated was growing on Gunung Beramban, Cameron Highlands.
Photo: B Parris

to find a rare species represented by one dead frond in an unsorted mat of material sent to Kew, which could have been identified and more gathered had it been dealt with properly in the field!

Medium-sized ferns, those where one frond fits into a drying paper, are usually the best represented, although quite often there is no rhizome material or information as to whether it is erect or creeping and, if creeping, how widely apart the fronds are produced. Large ferns, those where it is impractical to collect the whole frond, are usually poorly represented in herbaria. Often the apex of the frond or

Dipteris novoguineensis grows in open situations on mountains from Peninsular Malaysia to New Guinea. The young fronds are tinged red, but the colouration disappears as they mature. This plant was photographed on the summit trail of Mt Kinabalu above Kamborangah, Sabah.
Photo: B Parris

one or two pairs of segments (pinnae) are taken, but these fragments are rarely adequate for the identification of difficult groups. Yet it is relatively simple to make good specimens which can be mounted using only three herbarium sheets. The base of the stalk, including the scales of the rhizome, should be collected, as should the base of the frond, its central portion and its apex. The dimensions of the frond should be noted, as should the habit of the rhizome, whether erect or creeping.

Although fern collecting differs in some particulars from flowering-plant collecting, the methods are sufficiently similar for the fern botanist to work with flowering-plant botanists on expeditions. Essentially they share the same problems of drying collections or preserving them in alcohol, except that the smallest and most delicate ferns, such as the Hymenophyllaceae, with the blade of the frond only one cell thick, can be dried quickly and easily in a small lightweight metal frame-press over a camp fire or pressure lamp. Separating these little ferns from the rest of the collection pays dividends as they are not damaged by being inadvertently placed next to a much bulkier plant in a bundle or press of specimens.

It is very noticeable that in the field the fern specialists tend to work more slowly than the flowering-plant collectors. The latter will be searching for the distinctive colours of flowers or fruits whereas the former will be looking for differences in the shape of the foliage and the more subtle differences in shades of green to detect the fronds of ferns amongst the other vegetation of the forest. The fern collector is likely to spend much time closely examining fallen trees to gather ferns growing on their branches and is especially attracted to streams and riverbanks, where the high humidity and increased light favour the growth of ferns. Equally, high ground such as ridge tops, which may attract cloud cover for part of the day and thus have increased humidity and greater precipitation, encourage the growth of ferns.

Flowering plants are gathered when in flower or fruit, and flowering in particular can be a short-lived condition dependent upon rainfall or other climatic or seasonal variables, so the flowering-plant collector must be aware of all such factors in the areas where he or she wishes to collect. In contrast, the fern collector is able to collect essentially all year round in the wet tropics because, although there may be some seasonality in the production of fern fronds, the fertile parts stay in good condition for some considerable time. Even in conditions of local drought it is possible to revive wilted fern fronds by soaking them in water for several hours until they become good herbarium specimens.

Ideally, fern material to be collected will be mature, that is, with ripe spores, and these can be detected in the field with a little practice. A × 10 hand lens should always be carried when collecting and with its aid the dark sporangia or sacs full of spores can be seen. However, even if the spores appear to be shed, the frond can still be useful as a specimen. A few spores can usually be found in such material during microscopic examination back in the laboratory, and even without these the important distinguishing characters, such as presence or absence of an indusium, the protective covering over the sporangia,

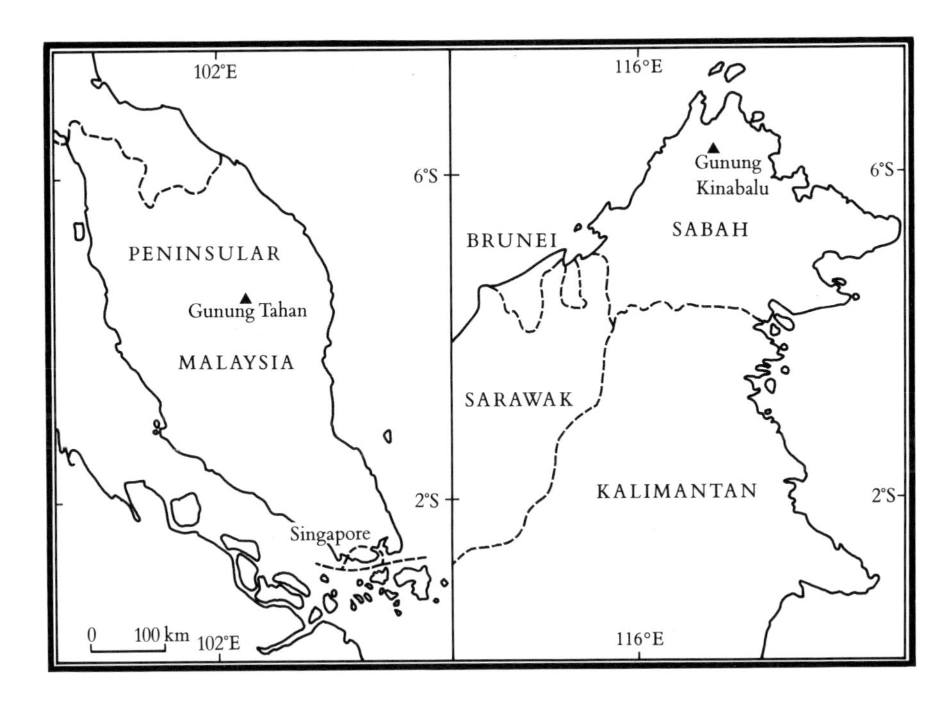

Map 16 Peninsular Malaysia and Sabah with places mentioned in ch. 16.

which are usually aggregated into a cluster (sorus), and the distribution of these clusters, can be seen.

Fern expeditions to Malaysia and New Guinea

The specialist fern-collecting expeditions from Kew are usually to tropical rainforest in Southeast Asia and South America because these forests are rich in ferns and underexplored botanically. Lowland rainforest in particular is quite undercollected, even in botanically relatively well-known areas. For example, Kew expeditions in 1985 to Taman Negara in Peninsular Malaysia made some surprising discoveries. Previous collectors, including a fern specialist, had climbed Gunung Tahan, the highest mountain in Peninsular Malaysia, but none had explored the rich lowland forest at its base. The two Kew collectors recorded three genera and 39 species new to Taman Negara, together with two species new to science, during nine days of fieldwork.

At 4100 metres (13,450 ft) Mt Kinabalu, in Sabah, Borneo, is the highest mountain in Southeast Asia (outside New Guinea) and also the best known botanically, although most collecting has taken place on the main summit trail on the south side and, apart from the Marai Parai spur on the west and the eastern shoulder, it remains virtually unknown. Even the main trail, however, has surprises for the specialist fern collector on the lookout for a particular group.

On one of my visits to Kinabalu searching for members of the Grammitidaceae, a family of small usually epiphytic ferns characteristic of and important in montane tropical rainforest with many species superficially closely resembling each other, I found five species new to Kinabalu and 14 undescribed species, of which eight had never been collected previously. That is a good indication of the importance of the

This stag horn fern, *Platycerium coronarium*, was photographed on Penang Island, Peninsular Malaysia. It is quite common as an epiphyte in lowland rainforest from Burma to the Philippines and is also cultivated in Europe, Australia and North America. The upright sterile fronds form a basket into which leaf-litter falls and decays, providing food for the fern. Rainwater is also trapped in the basket, which acts as a reservoir. The fertile fronds are pendulous.
Photo: B Parris

trained collector's eye in finding species which had doubtless been seen by other botanists but had not been recognised as distinct from those they had already collected.

Borneo, the second largest tropical island in the world, is an ancient and important centre of fern diversity in the Old World and as such is important for research at Kew. East of Borneo lies New Guinea, the world's largest tropical island, which is estimated to contain some 2000 fern species, twice the number of Borneo and about a sixth of the world's fern flora. Exciting fern discoveries can be made there with very little effort; in fact collecting along the roadsides, a habit scorned by many flowering-plant botanists, has produced species new to science, and forests easily accessible by road contain a rich assortment of ferns.

Perhaps the most interesting fern communities in New Guinea are the tree ferns (*Cyathea*) in alpine and subalpine grassland which support a number of epiphytes, both ferns and flowering plants, on their stout trunks. The richest development of these communities is on the higher mountains where there is no road access, and those who wish to collect there must hire a guide from the nearest village, shoulder a pack and be prepared to walk all day along a foot track which will meander through muddy pools, along slippery tree trunks and across streams which with luck will not be in flood with the almost inevitable afternoon rain until the 'campsite' is reached.

This is usually a sloping piece of ground at or near the tree line, always boggy and often without a supply of water other than rain, free enough of fallen trees to pitch a tent but without other amenities and as

Campsite at the margin of a grove of tree ferns (*Cyathea percrassa* and an undescribed species of *Cyathea*) at c 2800 m (9200 ft), near the summit of Gunung Binaia, a limestone mountain in Manusela National Park, Seram eastern Indonesia, which was of great interest to scientists participating in the Operation Raleigh expedition to Seram in 1987. In the foreground is a mixed colony of bracken (*Pteridium aquilinum* var. *wightianum*) and *Histiopteris sqamulata*, a relative of bracken which, like *Cyathea percrassa*, is known to occur only in Seram and New Guinea. The ridge in the background is covered with bracken and the low turf between supports colonies of the adder's-tongue fern (*Ophioglossum nudicaule*).
Photo: Alpics

cold, wet and foggy in the later afternoon as an English winter's day. After changing into dry clothes, preparing a hot drink and then a meal on a tiny methylated-spirit stove, crawling into a sleeping bag, night will have fallen and there is nothing else to do but to try and sleep through the activities of small mammals scurrying around and over one's body and the sound of rain on the tent.

Morning is different. The sneaking feeling of the night before – that there may be better places to be and better jobs to do – disappears. In brilliant sunshine under a cloudless sky it looks a different world. Uphill are the bare mountain peaks, downhill beyond the edge of the forest are range upon range of lesser peaks and all around are the tree ferns in golden grassland. Brilliantly coloured orchids (*Dendrobium* section *Oxyglossum* in particular) are in flower and birds of paradise fly overhead.

The day is spent collecting as the morning sun gives way to cloud, followed rapidly by rain; then the specimens are brought back to the tent to be prepared for the return walk the next day. After another disturbed night the pack, heavier now with the load of plants, is shouldered and the path to the nearest road is retraced. Then one gets back to civilisation by whatever means of transport is available – it has ranged from a forestry department Land-Cruiser or a missionary truck to a local PMV (public motor vehicle) with village people wearing headresses of bird plumes – to press the plants and to change into clean dry clothes.

At lower altitudes in New Guinea it is possible to undertake collecting without spending the night under canvas. The best days fern hunting in the lowlands and foothills are often in the company of young village children, who have a natural taxonomic talent and understand very quickly what is wanted in the way of ferns. They seldom collect the same species twice and each is aware of what the others are collecting, so it is possible to gather a lot of good material in a short time.

Multidisciplinary expeditions

Not all fern collecting is as arduous as that in the highlands of New Guinea, however. On Mt Kinabalu the main trail has good accommodation for the many tourists who wish to climb to the summit, and botanists wanting to study there are able to take two days over the journey, staying overnight in huts with comfortable bunks and working after dark. In terms of productive fieldwork there is much to be said for fern collectors joining expeditions arranged by organisations such as the Royal Geographical Society and the young people's Operation Raleigh, who require botanists as members of a team of scientists. Often it is easier and quicker for them to arrange visas, travel, accommodation, food and porters than it is for individuals, and on a large expedition there is usually the added benefit of a field laboratory with enough bench space to sort plants properly and an electricity

The glaucous fronds of the epiphytic filmy fern (*Pleuromanes pallidum*) are a characteristic feature of mountain rainforests from Ceylon to Polynesia. This fine colony was photographed on Gunung Beremban, Cameron Highlands, Peninsular Malaysia.
Photo: B Parris

generator to enable this to be done in the evenings so that the daylight hours can be spent in fieldwork.

It is useful too for fern specialists to discuss their work with other scientists. For example, it is interesting to have identifications of the host trees upon which epiphytic ferns grow to see what, if any, ferns are restricted to a particular family, genus or species of host, and it is useful for entomologists to have a fern expert to hand when identifications of ferns on which insects feed are needed.

Recently Kew fern botanists have participated in several multidisciplinary expeditions with great success. The work undertaken has included not just the basic collecting with good field notes which provides the foundation for research in the Herbarium, but more detailed work on fern distribution, such as those adapted to streambeds (rheophytes), the zonation of fern species on mountains, their diversity in tropical rainforest in terms of the number of species in a given area and surveys of host – fern epiphyte associations. Such studies provide information important to the classification of ferns and will probably become an integral part of expeditionary fern studies in the future.

17 *Fungi – the Fifth Kingdom*

DAVID PEGLER

Fungi comprise a kingdom of organisms showing an almost infinite diversity in form and variety. The number of species described as new to science currently runs in excess of 1500 per year. They are very important, not only as agents of decay and organic breakdown, but also in their essential association with the roots of many plants, including herbs and especially trees.

In addition, their use to mankind either directly as a food source or indirectly as the raw material in the biotechnological manufacture of so many increasingly useful products should not be underestimated. The vast majority of 'new' fungi described continue to come from the temperate and boreal regions of the world, and this is largely because the tropical and subtropical zones remain mycologically unexplored. This is one of the main reasons why organised expeditions especially designed to collect fungi are so essential.

Of course, not all mycologists come under the heading of 'mushroom hunters'. Merely bending down to pick up a handful of dead and rotting leaves from the forest floor can provide enough research material to keep an expert on the two groups of fungi *Hyphomycetes* and *Coelomycetes* busy for years. But, for instance, the fungi which cause stem and leaf diseases need to be collected together with their host plants, and specialists of these groups must be equipped with the apparatus and expertise usually more associated with the flowering-plant botanist.

Some fungal experts specialise in aquatic fungi, either marine or freshwater, and will devote their attention to sunken leaves and floating driftwood. There are specialists interested only in subterranean (or hypogeal) fungi, who may require extra assistance, such as the sense of smell demonstrated by a dog, or keen observation of the scratchings around trees made by wild birds and animals. Finally, there are the fungi familiar to everybody, collectively called 'Macrofungi', which

The tropical South American agaric *Inopilus virescens* with beautiful blue-violet gills. The colour is only seen when collected alive in the field as it is lost when the specimen is dried.
Photo: D Pegler

include the mushrooms and toadstools, the bracket fungi, club fungi, puffballs and stinkhorns, and the cup- and flask-fungi. These larger fungi may be either on the forest floor, perhaps half-buried and almost totally obscured by deep humus or leaf litter, in open grassland; or they may grow on wood which may be either dead and rotting on the ground or part of a living tree so that fruitbodies can be found up the side of a trunk or higher up on spreading branches. Fungus collectors may have to become expert tree-climbers or, as is recorded in one instance, train a monkey to do the job.

The difficulties of collecting fungi

Obviously, collecting each of these fungal groups requires specialist techniques and careful searching. On any expedition, it is the macro-fungi which, although the most easily observed, can often prove to be by far the most difficult to collect.

Fungi have very precise seasons yet are, for the most part, very short lived, perhaps lasting only a few hours in any one year. When launching an expedition, it is essential to know when to collect. In a temperate country, the season can be easily defined. The major 'fungus season' occurs throughout the months of autumn after the hidden, underground threads (mycelium) of each fungus species has experienced during the summer months the optimal conditions for growth, and when the flowering plants, both herbs and trees, are beginning to decline, so reducing competition for the fungus. In tropical climates, the seasons may not be so well defined but the fungus season will generally coincide with the monsoon or rainy season(s). Usually the best collecting is experienced during the early weeks of the rains before the heavy downpour finally washes the mycelium out of the soil.

(*opposite page*) A watercolour of *Pholiota brittonae* by Dr R W G Dennis, who collected this fungus in Trinidad and was for many years a mycologist at Kew. Such field sketches, like colour photos, are a valuable record to supplement the dried specimens.

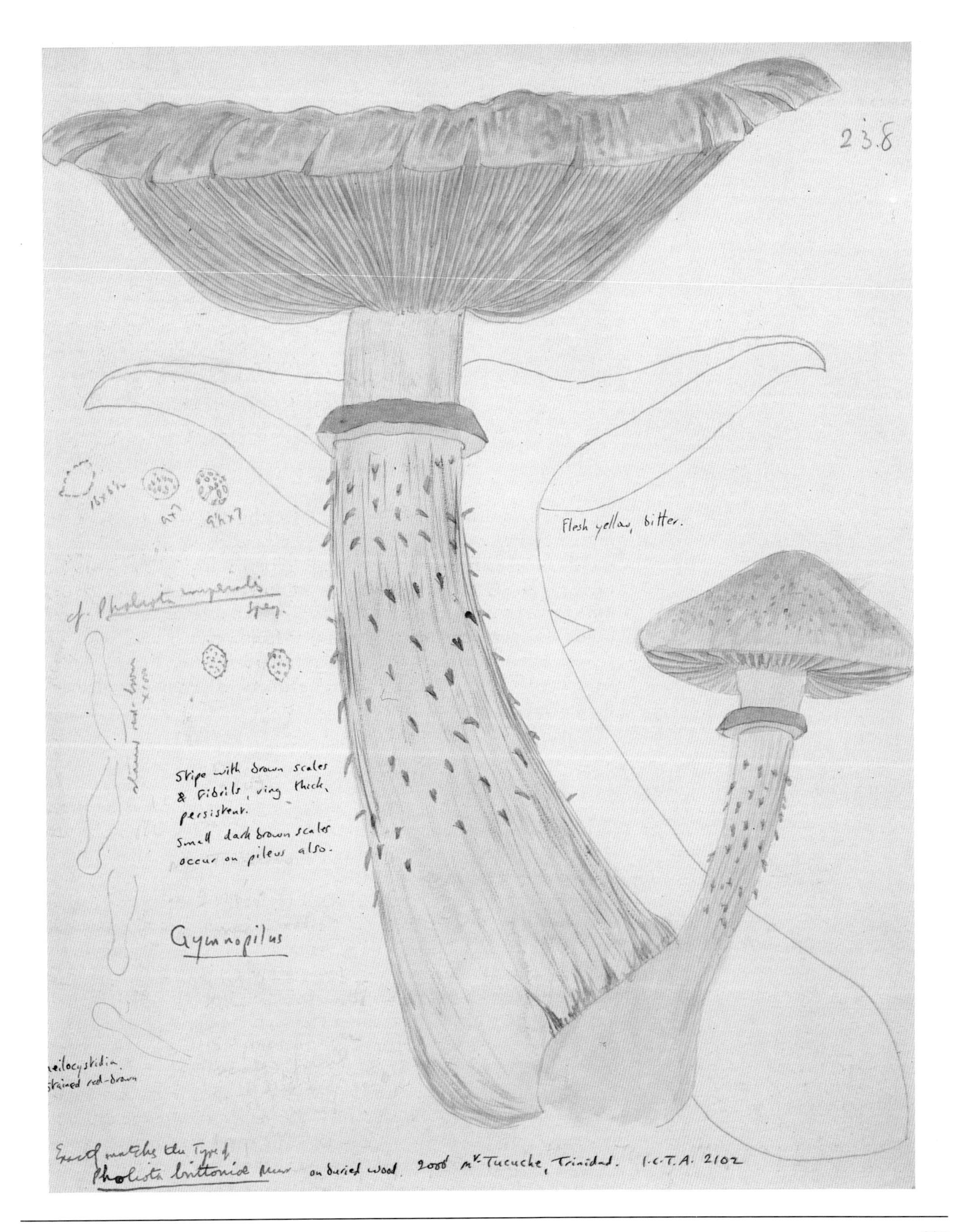
23.8
Flesh yellow, bitter.
Stipe with brown scales & fibrils, ring thick, persistent.
Small dark brown scales occur on pileus also.
Gymnopilus
eilocystidia
stained red-brown
Exactly matches the Type of
Pholiota brittoniae Murr
on buried wood. 2000' Mt. Tucuche, Trinidad. I.C.T.A. 2102

Collecting a bracket fungus, *Piptoporus* species, from a fallen baobab tree in the Kimberley region of the northern part of Western Australia, May 1988.
Photo: B M Spooner

Many species have a relationship confined to a particular flowering-plant species, so before the fungus is picked the collector must make a note of the habitat and the substratum, that is to say he or she must have a knowledge of the flowering-plant species with which the fungus is associated. For example, the obligate mycorrhizal fungi, which can only live in association with tree roots, are found only in the appropriate growing conditions. For many years it was assumed the genera of the agaric family Russulaceae, namely *Lactarius* and *Russula*, were absent from the tropics, and indeed they are, for the most part, from the tropical rainforest. When moist evergreen forests on mountain slopes or dry-type forests near the coast are explored, however, then members of Russulaceae may be found in abundance. It is therefore worthwhile appreciating that the energy-expensive mycorrhizal tree species cannot compete in the optimal growing conditions of the tropical rainforest, yet in conditions offering only restricted growth, such as poor soil and hard rock or low rainfall, mycorrhizal associates come into their own.

The next problem for the expedition member is to find a range of fruitbodies for collection. This is particularly true for the mushrooms and toadstools, collectively known as 'agarics' (order *Agaricales*). A fruitbody can change enormously during its very short life, not only in form and size but often in colour. Many species of *Cortinarius*, for instance, will have deep purplish or violet-tinged colours in the very early stages which can totally disappear by maturity. Without this information, the taxonomist cannot hope to provide an identification. Similarly, structures such as the veil can be weathered away as the fruitbody ages, or the initially slimy surface of the cap may dry in the sun and become cracked. The problem here is that tropical agarics, unlike most temperate species, tend not to form large troops but occur only as solitary individuals, so he may find only one poor specimen.

How to collect fungi

It is essential to collect only fruitbodies in good condition; preserving them in that state is usually accomplished by using a flat-bottomed basket to minimise damage. Small specimens are often best gathered by placing them in individual containers, and metal tobacco-tins have proved most effective for this purpose. Fruitbodies can deteriorate extremely rapidly, especially in hot and humid conditions, and some personal success has been achieved by taking a picnic-style icebox, complete with ice-cubes, into the forest in order to transport specimens satisfactorily back to the laboratory.

Fruitbodies need to be collected in their entirety: all too often the base of the stem is left behind by inexperienced collectors. In many of the tropical countries of the Old World, there occurs a group of mushrooms called *Termitomyces*. Their fruitbodies originate from the knots of fungal threads in an active termite comb, which may be found either within a termite-mound or buried deeply underground. When one of these mushrooms is gathered, it may become necessary to dig into the rock-hard lateritic soil, perhaps for a metre or more in order to ensure that the long root-like base (or pseudorhiza) is collected with the fruitbody.

Most soft, fleshy fungi deteriorate noticeably in appearance once an attempt is made to dry or otherwise preserve them, the properties of colour and shape being entirely lost. It is therefore absolutely necessary to make full descriptive notes of the material collected in the field. This should extend far beyond the recording of the size and form of fruitbodies. Colour is of paramount importance, including the change of colour that can occur through oxidation processes, once the flesh is broken open and exposed to the air. The presence of any exudate (latex) needs to be noted, as well as the characteristics of odour and taste.

In the forests of the Caribbean, for example two frequently occurring extremely similar small white agarics, *Camarophyllus niveicolor* and *C. buccinulus*, can be readily separated by the characteristic aromatic aroma of peppermint emitted by the former. In the Ugandan forests, a third species, *C. olidus*, has a very pungent farinaceous smell, which can be detected at a distance of several metres. Details such as this are completely lost once the specimens have been dried for preservation. It is always important to try to obtain some kind of illustration, either in the form of a watercolour sketch or a photograph.

Another extremely useful technique when collecting larger fungi is the making of a spore-print. Many tropical bracket-fungi exhibit an extremely short sporulating period, perhaps for no more than 24 hours in any one season so that the vast majority of fruitbodies, upon examination, are found to be sterile. There is therefore an urgent need to acquire descriptive information concerning the spores and the spore-producing structures. Spore-prints are made by placing the lower surface facing downwards onto a sheet of paper or a glass-microscope slide. Spores deposited as a 'print' have been actively released by the fungus and are therefore mature. They can remain viable for at least several days, and freshly made spore-prints can be sent to a suitable laboratory so that the fungus may be grown in culture.

Termitomyces eurrhizus, from equatorial Africa, has a false root (the pseudorhiza), which is normally hidden underground and rarely collected. A mycologist such as Dr David Pegler, who knows this fungus, takes great trouble to unearth it.

How to preserve fungi

Fungal fruitbodies gathered on an expedition need to be preserved before they can be transported. Unlike flowering plants, the fleshy fungi must never be pressed. They lack both a vascular and a cellular structure and will be destroyed by a plant press. Instead, fruitbodies must be carefully dried either in their entirety or, in the case of large woody specimens, as slices.

The process of drying is critical for the success of the preservation. Too much heat will only succeed in first cooking and then charring the fruitbody, whereas too little heat will only result in hatching out any insect larvae present, which will rapidly eat the specimens. A closed oven is not suitable; what is requested is a system involving a continuous stream of warm air, perhaps by using a fan-heater when electricity is available. In the field, it will be necessary to find an alternative heat-source, such as a charcoal fire, an oil-lamp, a paraffin stove or a gas-cylinder. As last resort, air-drying in the sun may be attempted but this is rarely successful, except for very small specimens.

Tropical climates introduce an additional hazard for the fungus collector, namely high humidity. A dried fruitbody must be kept dry for it tends to be hygroscopic, quickly absorbing any moisture in the air, and once damp it will provide a perfect substrate for rapidly-growing moulds so that the specimen will be destroyed. Problems of storage are normally overcome either by placement in a desiccator over calcium chloride or by sealing in polythene bags, preferably with a few granules of silica gel. It is worthwhile placing several smaller bags within a larger clip-top bag.

An alternative means of preservation is to place the collections in a preservative spirit, although this has distinct disadvantages for the taxonomist who will eventually wish to examine the material. Spirit has a tendency to leach out any pigments present, and also may render any chemical stain-reaction ineffective. The procedure can be useful, however, for certain larger specimens or for delicate material, such as the stinkhorn fungi and their relatives (order Phallales). When such material is despatched, both bulk and weight can be considerably reduced by wrapping the material in impregnated cotton-wool, sealed in a polythene bag.

★ ★ ★

The mycologists at Kew have world-wide field experience and they publish their findings for the benefit of science but much remains to be done. Expeditions designed to collect fungi are desperately required. Much of equatorial Africa, South America and Southeast Asia remains totally unknown, even though survival of the forests is dependent upon these organisms.

18 Plant collecting and conservation

GRENVILLE LUCAS

Brazilian rainforest burning on one of the big cattle-farming projects on the Trans-Amazon Highway near to Altamira. There is great international concern at the destruction of the Amazonian forest with consequent loss of its plants and animals – the irreplaceable genetic resources. In 1988 alone, an area the size of Belgium was destroyed.
Photo: G T Prance

The mid-18th century was a time of great scientific discovery in Europe and Britain's monarch, George III, encouraged many scientists to seek knowledge for its own sake. A major driving force behind him was Sir Joseph Banks, a man who has many claims to fame, not least as president of the Royal Society. Banks saw the exploitation of the world's living natural resources as a method by which Britain could develop her

empire. His own voyage of discovery (1768–71) with Captain James Cook gave a powerful, dual thrust to some of his writings: he recognised the need to explore the unknown for its own sake. On the one hand there were the vast numbers of new species waiting to be found and catalogued, and on the other the potential value of 'new' products both in local trade and on a world scale.

The previous chapters have demonstrated that these two related themes of 'pure' research and commercial exploitation are as fundamental to Kew today as they were in 1773, when George III and Banks, as his adviser, set about making the royal gardens at Kew a centre for many plant wonders from around the world. However, the emphasis has changed dramatically, particularly in the last 10 to 15 years as the awareness of man's complete dependency on plants, and just what that means, has grown. The conservation of plants is a complex and many-faceted problem and this chapter deals only with those elements that relate to exploration.

There are the general topics. For example, we need to know what the flora of the world is, and what percentage is threatened; where does it occur and what are the threats? A European and North American perspective – let us call it the developed world's perspective – on this problem is quite misleading. Here the botanical community has a reasonably comprehensive knowledge of where each species is to be found. For instance, in the vast majority of cases, and backed up with much detailed information, we can pinpoint all United Kingdom species to a 10-km or even 2 km grid square and all the rare species can be located to their exact site. This data not only identifies the species, but in most instances the size of the population which is vital in order to ensure the long-term maintenance of genetic variability of the species.

So there is, at least in some parts of the world, a database indicating which species are threatened and what the threats are. Add to this the ability to protect particular habitat types and the problem is half solved. The other half of the solution is quite simply to have the resources and willpower to reverse the situation. In the developed world there is increasingly a sufficiently high agricultural production level to enable plans to be developed for the integration of agriculture and conservation in each country and the finance to put these plans into action. That is not to say that such plans exist in many cases.

A most important aspect of expedition policies of institutes such as Kew is to locate and verify the plant rarities and to develop the techniques whereby living material, preferably in the form of seed or failing that as cuttings, can be propagated to increase breeding populations of endangered species. At the same time it is important to select the widest possible genetic variation to ensure both the maximum diversity and the long-term survivial of the species and to allow the slow processes of evolution in 'the wild' to continue.

It is easy enough to describe the solutions, but many years of field and laboratory study are probably needed for each species. In the British context, with a total flora of approximately 1700 species and under 200 in need of in-depth research to ensure their survival, this may be feasible. But if we look at the problem on the European scale, we face a total flora of some 14,000 species, with 2–3000 in need of this in-depth approach. A similar scale of problem exists in North America, but it must be multiplied many times over in the tropics. Therefore, expeditions and allied research have to be focused on those groups which have been identified as having particular merit or use i.e. plants which are horticulturally attractive, scientifically interesting or of potential economic or medicinal value.

The popular belief that once we know the name of a plant we know all about its biology is one which constantly surprises the staff at Kew. It is, no doubt, flattering to the scientist, but it is far from the truth and potentially a very dangerous misconception. Plant species are not isolated entities in their own right – they are part of a complex integrated ecosystem. The fungi in the soil that 'help' the seeds germinate, absorb water and mineral salts from the soil, even protect them from other destructive fungi; the ants which appear to 'clean' plant leaves in return for nectar and a home; the pollinating insects and seed-dispersal animals – all are part of a dynamic and integrated system. Conserving a few specimens of a species may not actually help to save an ecosystem. That is why the whole biology of any particular species must be studied by those on field expeditions; this data can be vital for any reintroduction to the 'wild' – even a wild altered or interfered with by man.

Today, the many specimens obtained on expeditions are deposited in the host partner's national or regional herbarium, with a duplicate set kept at Kew. In the past these collections were often made with no regard to the conservation status of the species in the wild, even when it was known to be rare. Nowadays, every care is taken to ensure the population remains secure in the wild.

Here, however, is one of the saddest problems that botanical science faces today, because in the past, when a new species was described, its exact location was openly published. Now we have to lock away many rare, beautiful and interesting species so that the unscrupulous 'collector' cannot, having discovered the exact locality, dig up the rarity from the wild for sale to the highest bidder, with the result that whole wild populations may be eradicated. Groups such as the succulents, particularly cacti and aloes, and some orchids, have been driven to extinction by such greedy people, who have used the very discipline of good scientists against the plants they once sought for study. To combat such practices, Kew was one of the advisers at the 1973 Washington meeting to draw up the Convention on International Trade in Endangered Species (CITES), which seeks to ensure that conservation principles are turned into international and national laws to safeguard wild populations of species from unscrupulous traders.

The field notes giving local names, altitude, habitat, soil type, general ecology and uses by local peoples have been the starting point for all the major *Floras* Kew has produced over the last 150 years. Such work represents Kew's constant commitment to the tropics and their conservation and development, but it also serves to expose how little

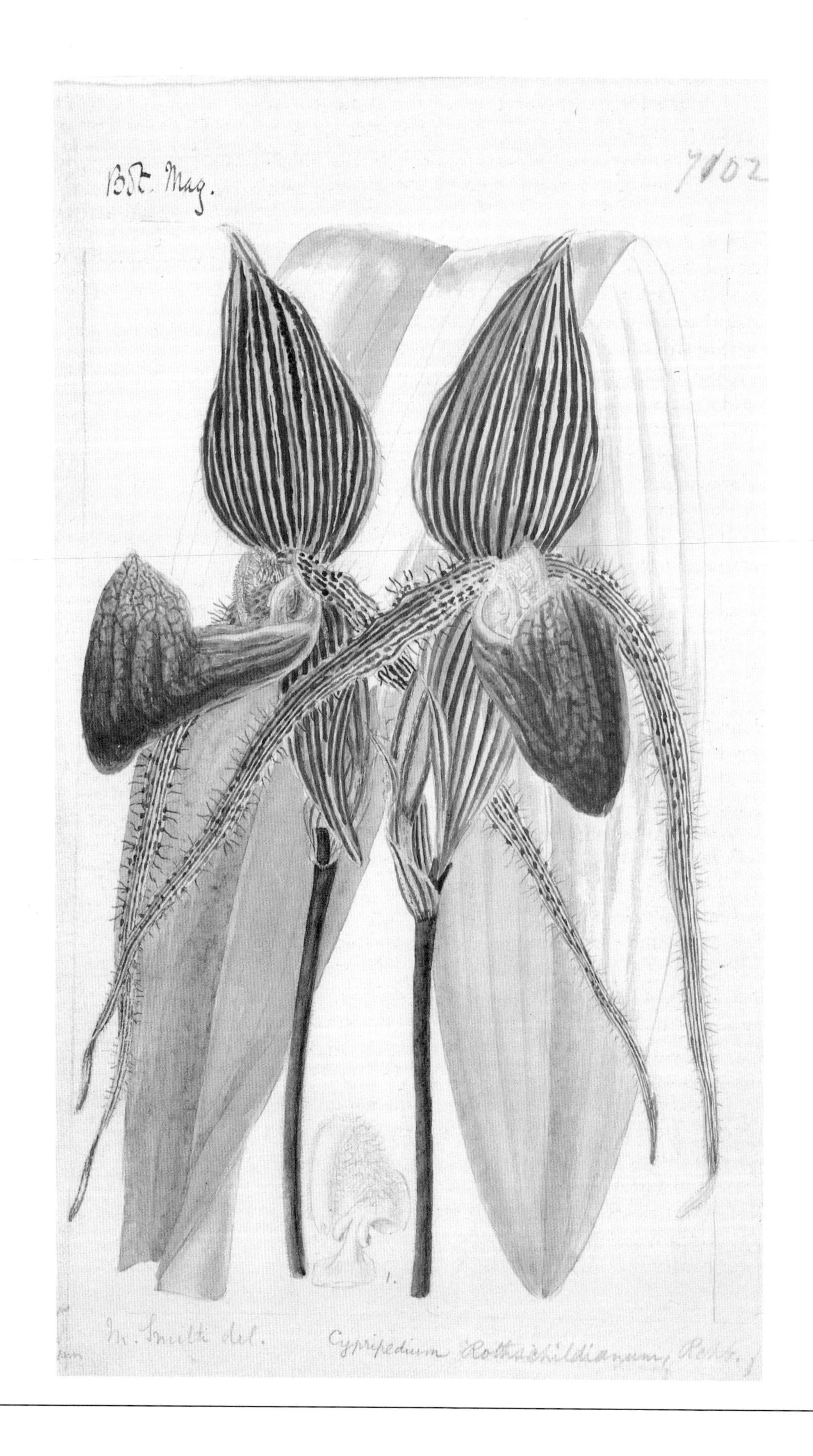
Bot. Mag.
7102
1.
M. Smith del.
Cypripedium Rothschildianum,

we really know about the southern hemisphere – another and much more important part of our expedition story. Whereas, as mentioned earlier, we have at least a reasonable knowledge of the vast majority of species' distributions in the developed world, we have a frighteningly poor knowledge of what are probably common species in many tropical countries.

This is, in many cases, a simple physical problem: a limited number of botanists working in huge areas. Australia provides an excellent example. It has a most active set of state and federal botanic gardens and a fine, but small, band of active botanists. But the sheer size of the land mass means that there are still many hills and valleys in remote regions to explore botanically, where the climatic stresses and logistical support needed provide the same personal tests for initiative and tenacity that the very earliest explorers faced. This element of adventure, and the development of personal reliability and awareness, strengthens the character of all expedition members and this, in itself, reflects upon their parent institutions.

This book shows varied facets of its contributors and their expedition targets and successes. However, there is one very clear consensus – the rapidly growing concern over the degradation and loss of vegetation cover, habitats and species. Alongside this devastation has come the attendant increase in what are often claimed to be 'natural', often local disasters – flash-floods and mud slides and the consequent loss of what was formerly good agricultural land. One reads and sees references to these problems, in the press and on television, but to see them for oneself, to live with the people in these regions, makes one realise all too forcibly what is actually happening. Man is steadily destroying 'his habitat' and his future.

The thrust, therefore, of all Kew's tropical expeditions, whether to the arid or humid regions, is now more strongly than ever stimulated by the need, not just to identify the flora, but to see what species are used and for what purpose. We do this in pursuit of our basic belief that the answer to tropical problems will be found in tropical solutions.

The fundamental work of exploration at Kew continues with partners in many countries, as the preceding chapters show. *Floras*, checklists and guides will remain a basic element of our work, but the need to enlarge our potential economic-plant research side, in relation to the expedition programme, has seen the wheel come full circle.

Plants and their products can solve all man's food, fuel and fodder needs. Our databases, built on the toil of many expeditions and their collections, hold the key to these problems. It is the political willpower that is needed. The bilateral and multilateral aid agencies are beginning to see that perhaps the solutions are relatively modest in financial terms, but vastly complex to administer. If man is to succeed, our knowledge must be turned into agricultural reality.

(*opposite page*) This slipper orchid *Paphiopedilium rothschildianum* is one of the rarest plants in the world. Although protected to the extent that it grows only in the Kinabalu National Park, Sabah, it is collected to near extinction by unscrupulous orchid dealers. Original drawing by Matilda Smith for *Curtis's Botanical Magazine*, plate 7102 (1890).

Firewood gathering, such as here near Annapurna in Nepal, can hasten deforestation and soil erosion. Controlled collection at a sustainable rate maintains the natural woods; replanting with alien species such as eucalyptus often eliminates the native species.
Photo: A D Schilling

Further Reading

ALLEN, Mea *The Hookers of Kew 1785–1911*. London, Michael Joseph, 1967.

BAKER, J.N.L. *A History of Geographial Discovery and Exploration*. New York, Cooper Square Publishers, 1967.

BEAN, William J. *The Royal Botanic Gardens, Kew; Historical and Descriptive*. London, Cassell, 1908.

BROSSE, Jacques *Great Voyages of Discovery: Circumnavigators and Scientists 1764–1843*. New York and Oxford, Facts on File Publications, 1983.

COATS, Alice M. *Quest for Plants: A History of Horticultural Explorers*. London, Studio Vista, 1969/New York, McGraw Hill, 1970 (USA title: *The Plant Hunters*).

COX, E.G. *A Reference Guide to the Literature of Travel*. New York, Greenwood Press, 1969.

DESMOND, Ray *Dictionary of British and Irish Botanists and Horticulturists*. London, Taylor & Francis Ltd, 1977.

HEPPER, F. Nigel (ed.) *Kew: Gardens for Science and Pleasure*. London, HMSO 1982.

HUTCHINSON, John *A Botanist in Southern Africa*. London, P.R. Gawthorn Ltd, 1946.

LANCASTER, Roy *Travels in China: A Plantsman's Paradise*. Woodbridge, Suffolk, Antique Collectors Club, 1989.

LEMMON, Kenneth *The Golden Age of Plant Hunters*. London, Phoenix House, 1968.

LYTE, C. *Sir Joseph Banks: 18th Century Explorer, Botanist and Entrepreneur*.

TURRILL, W.B. *The Royal Botanic Gardens, Kew, Past and Present*. London, Herbert Jenkins, 1959.

TURRILL, W.B. *Joseph Dalton Hooker: Botanist, Explorer and Administrator*. London, Nelson, 1963.

Index

NB: Page numbers in italics refer to illustrations.
Numbers in bold are map numbers.

A

Abies delavayi 97, 100
Abutilon striatum 167
Acacia 30, 148
Acacia ehrenbergiana 75
Acacia tortilis 75
Acantholimon diapensoides 91
Acer forrestii 100, 101
Acer giraldii 101
Achillea cretica 82
Acosmium tomentellum 158
Actaeon, HMS 14
adder's-tongue fern *201*
Adenium obesum 68
Adiantum reniforme 48
Aeolanthus 26
Afghanistan 85–91
  map **8**
Africa
  Hovercraft expedition 28–31
  Jodrell expeditions 183–6
  tropical 20
agarics 206, 207
AIDS research 155
Aiton W T 6, 7, 16
Albizia 113, 148, 160
Aldabra
  expedition 61–71
  map **6**
  mixed scrub 66
  South Island 66
  vascular plants 65
Alexa canaracunensis *152*, 155, 156, 158–9, 160–1
Alligator, HMS 11
Alnus subcordata 163
Aloe 183–5
  biochemistry 185–6
Aloe arborescens 44
Aloe ferox *184*
Aloe mawii 44
Aloe perryi *68*
Alpine Garden Society 174
Alstonia 113
Alyssum chondrogynum 84
Amherstia nobilis 119
Anadenanthera 148
Anatolia *81*
Andes collections 182–3
Andira 148
Androsace delavayi 100
Androsace rigida *101*
Anemone 86
Anemone blanda 83
Anemone tcherniaewii 88
Angil, Mount 114–16
Antirrhinum molle 167
Anton, William 1
Antongil, Bay of *53*
Antongilia perrieri 58
Apodytes 66
Apodytes dimidiata 67
Aqauria 44
Arabia, William Lunt's collections 19
Araucaria araucana 5
Araucaria spp 180
Arbutus menziesii 5
Arbutus unedo 82
Argentina
  Andes collections 182–3
  bulb and rhizome collections 178–80
  collections 16–7
  map **14**
Argyrolobium uniflorum 82
Arisaema handelii 102
Arisaema wilsonii 102
Armstrong, John 11, 12

Aroideae Maximilanae 133
arrow poison frogs *157*
Arum conophalloides 83
Ascoglossum calopterum *127*
Asparagus umbellatus *67*
Astelia alpina 110, *110*
Asystasia 70
Athyrium 196
Attaleinae 51
Attenborough, Sir David ix
Augusta, Princess 1
Australia
  Allan Cunningham's collections 8
  George Caley's collections 5–6
  Robert Brown's collections 4
  Martin Sand's collections *125*
  Brian Spooner's collections *206*
Azraq Oasis (Jordon) *80*

B

Bactris maraja 158
Bahia (Brazil)
  expedition 128–43
  map **12**
  seed bank collections 190
Bailey, F M 93
Bailey Hortorium 58, 177
balsam 44
bamboo, heavenly 7
Banda, Elias 42–3
Bandiagara Escarpment (Mali) *193*
Banks, Sir Joseph 1–2
  acquisitions for Kew 2
  collections sent to Kew 3–6, 7–8
  influence on George III 209
  plants of economic value 2–3
  selection of plant collectors 15
Barclay, George 11, 16
Barringtonia 55
Barter, Charles 14
Bartram, John 1
Bartram, William 1
Bauer, Ferdinand 79
Bauer, Fritz *37*, 38
Bauhinia guianensis 160
Bayer, Bruce 185
Bacquerelia clarkei 136
Begonia merrittii *125*
Bentham Moxon Trust 174
Berberis angulosa *92*
Berberis cretica *84*
Berlin National Garden Festival 175
Betula calcicola 100
Betula utilis 101
Betula utilis var prattii 100
Billbergia porteana *133*
Bligh, Captain William 3
Bolivia *141*
Bonn National Garden Festival 175
Borneo 200
Boswellia sacra *81*
Botanical (now *Kew*) *Magazine* xiii
Botrychium 196
Bougainville 119–22
Bounty, HMS 3
Bowie, James 7–8, 15, 16, 18
Brachystegia 38
Brandham, Peter 183
Brazil
  collections 183, 190
  expedition for useful legumes 149–61
  map of north-east **13**
  Maracá Rainforest Project 153–5
  rainforest *209*
  seed collection *188*
brazil-wood 149–53
Brazilian Academy of Sciences 129
breadfruit 3
Brewer, Captain 11
Brisbane River, Queensland *8–9*
Bromeliaceae, Andes collections 182–3
Brompton nursery 5
Brown, Robert 4, 6
Brownea 148
Brummitt, Hilary 42
Brummitt, R K, Malawi expedition 40–50
Buchanan-Hamilton, F 93
Buddleja saeviifolia 44
Bulbophyllum dennissii *117*, 118
Bulbophyllum hans-meyeri 115
Bulbophyllum medusae 120
Bulbostylis 70
Burke, Joseph 13
Burkill, I 93
Burton, David 4
Bute, Lord 1
butterwort 83

C

Caatinga thorn scrub 139, *142*
cacti
  Andes collections 182
  Brazilian *153*
Cadaba glandulosa 195
Caesalpinia 149–53
Caesalpinia echinata 148, 150, *151*, 152
Caesalpinia peltophoroides 148
Caesalpinia pulcherrima 148
Caesalpinia sappan 149
Calanthe cremeoviridis *109*, 110
Caley, George 5–6, 15
Calotropis 30
Camarophyllus buccinulus 207
Camarophyllus niveicolor 207
Camarophyllus olidus 207
Cameroon 28, 30
  expedition 1957 24–8
Campbell, John *126*
campo rupestre vegetation 137, 138
Canada, Archibald Menzies' collections 4–5
Cape-house (Kew) 2
carnaúba palms 139
Carter, Joan 89
Casabona, Isobel 179
Cassia aldabrensis 67
Cassia moschata *149*
Cassytha filiformis 67
castanospermine 155, 161
Castanospermum australe 155
Casuarina forest 63
Catalpa fargesii 97
Catasetum 135
cedar forest 83
Cedrus libani ssp *brevifolia* 78–9
Central African Republic 28
Centre for Economic Botany 147
Cephalotaxus fortunei 101
Cercis yunnanensis 98
Cereus hexagonus *153*, 154
Chad, Lake 28, 29, 30
Chamberlain, David 98
Chapada Diamantina *138*
Chambers, Sir William 1
Checklist of the Plants of the Seychelles 68
Chelsea Physic Garden 5
Chenopodium quinoa *149*
Chile, 1985 expedition 167–8
  map **14**
China
  Augustine Henry's collections 19
  Lichiang (Lijiang) region 99–101
  Mekong-Yangtse divide 101–2
  seed collecting *100*
  south western 96–102
  William Kerr's collections 6–7
  Yunnan Mountains map **9**
Chisenga *49*
Choiseul island (Solomon Islands) 123
Christensenia aesculifolia 113
chromosome evaluation 177
Cinq Cases camp *64*
Cistus *82*
Clathrotropis macrocarpa 160, 161
Clematis chrysocoma 98
Clematis hilariae 91
Clematis montana 100
Clinostigma collegarum 114

Cocoa Research Centre (CEPEC) 130, 131
Cocoeae 51
coconut, forest 51–60
Coelogyne veitchii 126
Coelomycetes 203
Collinson, Peter 1
Combretum 27
Combretum brassiciforme 27
Commelinaceae 177
Commissioners of Woods and Forests 11
Comoro Blue Pigeon 66
Conocarpus lancifolius *189*, 190
conservation of plants 209 13
Convention on International Trade in Endangered Species (CITES) 211
Cook, Captain James 1, 210
Copaifera 148
Cope, T A, Oman expedition 72–7
Copernicia prunifera 139
Cornus capitata 98
Cortinarius 206
Corydalis rutifolia 83, *84*
Corylus ferox 95
Cotoneaster cavei 95
Cotoneaster franchetii 97
Cribb, Phillip 117–27
Crocus cyprius 83, *84*
Crotalaria 26, 28
Ctenium 27
Cunningham, Allan 7–9, 15, 16, 18
Cupressus duclouxiana 98
Curtis, William 5
Cutler, David F
  African collections 183–6
  Andes collections 182–3
  Argentinian bulb and rhizome collections 178–80
Cyanotis 26
Cyathea 200
Cyathea gleichenoides 111
cycads 2
Cyclamen graecum 82
Cypella *180*
Cyperus papyrus 30
Cypripedium 100
Cyprus 78–84
  early collections 79
  map **7**
Cyprus cedar 78
cytological studies 177

D

Dacryocarpus compactus 110
Dalbergia spp 26, 148
Dalziel, J M 23
dandelion 84
Daphne bholua 164, *165*
Daphne bholua var *glacialis* 95
Daramola, Benjamin 26
Darian, Mandy 51, 52
Darwin, Charles 63
date-palms *76*
de Menzes, Dra Nanuza L 183
Delavay, Abbé 96, 98
Delonix regia 148
Delpinium brunonianum 91
Delphinium caseyi 83
Dendrobates leucomelas *157*
Dendrobium 201
Dendrobium decockii 111
Dendrobium gouldii 120
Dendrobium macranthum *120*
Dendrobium minutum 126
Dendrobium mohlianum 120
Dendrobium rhodostium 122
Dendrobium spectabile 120
Dendrobium vexillarius ssp *uncinatum* *115*, 116
Dennis, Geoffrey 118, 126
Dennis, Dr R W G *205*
Derby, Earl of 12
Derris 148
Deutzia monbeigii 101
Dhokal (Somalia) *175*
Dillenia 113
Dionysia 89, 90
Dionysia afghanica 90
Dionysia denticulata *89*
Dionysia lindbergii 90
Dionysia microphylla 90
Dionysia tapetodes 90
Dionysia visidula 90
Dipteris novoquineensis *197*
Discovery, HMS 4–5
Djam *87*
dos Santos, Talmon 133
Dracaena cinnabari *68*
Dransfield, John 51–60
Drosera 26
Drosera roraimae *140*
Drummond, R B 33
Dryolinnas cuvieri aldabrensis 71
Dryopteris 196
dyes 148
Dypsis humbertii 53, 58
Dypsis mocquerysiana 58

E

East Africa 31–40
  Colonial Office Expeditions map **3**
  Fourth Colonial Office Expedition 31–40
Ecuador 163–4
Eddy, Alan 104
Edinburgh Botanic Garden 98
Edwards, Peter *160*
Egypt 81
Elaeagnus multiflora *16*
Elaeagnus tricholepis 95
Elatine 82
Elburz Mountains 163, 166
Ellman, Revd E 20
Enarthrocarpus arcuatus 83
Encephalartos longifolius 2, 3
Endeavour, HMS 1
Enkianthus chinensis 102
Ensete gilletti 26
Entada polystachya *148*
Entandrophragma 47
epiphytes *182*
Eragrostis 70
Erebus, HMS 13
Eremurus spp 86
Eremurus stenophyllus *86*
Erica 44
Ericinella 44
Eriocaulon 26
Ernest Thornton-Smith Scholarship 173–4
Erskin, CM *166*
Erythrina 26
Erythrina crista-galli 148
Euonymus frigidus var *elongatus* 95
Euphorbia 70
Euphorbia abbottii 66
Euphorbia deightonii 30
Euphorbia griffithii 100
Euphorbia schillingii 95
Euphorbia sikkimensis *94*
Everest National Park *94*, 163
Evolvulus 70
expedition lorry (Kurawa to Malindi) *35*

F

Farges, Abbé 96
Farrer, Reginald 96
fennel, giant 88
ferns, tropical expeditions 196–202
Ferula 88
Fiji *125*
Fimbristylis 70
firewood collection *213*
Flacourtia 66
Fliegner, Hans 163, 166
Flinders, Matthew 4
Flora of the Arabian Peninsula 72

Flora Brasiliensis 139
Flora of Ceylon 68
Flora of Cyprus 72, 79
Flora Graeca 79
Flora Graecae Prodromus 79
Flora of Iraq 72
Flora of Mucugê 138
Flora Sinicae 98
Flora of Tropical Africa 23
Flora of Tropical East Africa 23
Flora of West Tropical Africa 23, 28
Flora Zambesiaca 23, 42
Floras 211, 213
Flore des Mascareignes 69
forage crops 190
Forgesia racemosa *69*
Forrest, George 96, 98
Fritillaria 86
Fuchsia loxensis 164
fungi 203–8
  collecting expedition *206*
  preservation 207–8
Furo de Maracá *152, 157*
Furse, Paul 86

G

Gaeryong National Park (South Korea) *166*
Gardner E 93
Gardner, George 13
Gasteria *185*
gene-banks 187
Genera Palmarum 51, 52
Gentiana douglasiana *5*
Gentiana falcata 91
George III, King 1, 209
Geranium papuanum 110
German Academic Exchange Scheme (DAAD) 174
  diet 66, 70
Gillett, J B *33*
Gibbon, Peter 177
Gladiolus 28, 44
Glandularia tenuifolia 167
Gleichenia vulcanica 110
Glionnetia sericea *68*
Gliricidia sepium 148
Gnidia chapmanii 45
Good, Peter 4
Goodia 4
Grammitidaceae 199
grasses of Aldabra *64*, 66, 70, 71
Grewia 195
Grewia salicifolia *67*
Grey-Wilson, C
  Afghanistan expedition 85–91
  Yunnan mountain expedition 96–102
Grove, Professor A T 28
Guadalcanal *121*, 126–7
Guaglianone, Rose 179
Guest, Evan 72
guides 213
Guiera senegalensis 195
Guppy, Henry 123
Gynerium sagittatum *141*

H

Halliwell, Brian 163
hallucinogens 148
Halmoorea 53
Halmoorea trispatha 58
Harley, Raymond 129, *160*
Harleyodendron 136
Harrison Church, Professor R 28
Hawkins, John 79
Haworthia collecting 183–5
heaths 2
Hedge, Ian 86
Hedychium coccineum 95
Heliamphora nutans *141*
Helichrysum 44
Helichrysum brassii *44*
Heliconia 123
Heliconia marginata *140*
Heliconiaceae 160
helicopters in plant collecting 77
Hemsley, J H 33
Henchie, Stewart 167–8
Henderson, Ruth 130
Henry, Augustine *19*, 96
Henry Idris Matthews Scholarship 174
Hepper, F Nigel 49
  Herbarium, Kew *145*
  West African expeditions 23–31
Herreria montevidensis 179
Hewer, Professor T F 87
Hill, John 1
Himalayas 93–5
  rhododendron collections *13*, 14
Himalayas, Joseph Hooker's collections 85
Hindu Kush *86*, 87
Histiopteris squamulata *201*
Hodgson, B H 93
Holmboe, Jens 79
Holmes, Susan 185, 186
Holt, Pamela *171*
Hooker, Joseph Dalton 13–14, 16, *68*, 93
  Himalayan collections 85
Hooker, Sir William 8, 11–12, 14, 18
  Niger Flora 23
Horn, Michael 193
horticultural plants
  contemporary collections 162–5
  new woody plants 162–6
Hortus Kewensis
  (John Hill 1768) 1
  (William Aiton 1789) 1
  (1810–13) 4
Hove, Anton Pantaleon 2, 16
hovercraft, plant collecting from *29*
Howard, Roger 166
Howard, Tony 177, 178
Hozelock Prize 174
Hsu Ting-zhi 98
Hubbard, Charles E 20
Hunt, David 177
Hunt, Peter 118
Hutchinson, John 19, 20, 23
Hymenaea 148
Hymenaea courbaril 148
Hymenocardia 26
Hymenophyllum 196
Hyparrhenia 27
Hypericum 26
Hypericum forrestii 97
Hypericum revolutum 44
Hyphaene coriacea *175*
Hyphomycetes 203
Hypoestes 70

I

Iguana iguana iguana *157*
Ilex yunnanensis 101
Impatiens 44, 57
Incarvillea lutea 100
Incarvillea olgae 91
Index Semina 187
India, Joseph Hooker's collections 13–14
Indigofera 26, 148
Indigofera pendula 100
Inga 148, 160
Inopilus virescens *204*
International Board for Plant Genetic Resources 187
International Institute for a Tropical Agriculture (Nigeria) 193
International Livestock Centre for Africa (ILCA) 190
Intsia 55
Investigator, HMS 4
Iran, Elburz Mountains 163, 166
Iris afghanica 88
Iris chryrographes 100
Iris heweri 88
Iris macroglossa 88

Iris pseudorossii 100
Iris purpureobractea *81*
Iris spp 86
Iris xanthochlora 88
ironwood, Persian 163, *164*
ixias 2
Ixiolirion montanum 88

J

Jade Dragon Snow Mountains *97*, 98
Japan, Charles Wilford's collections 14
Jasminum elegans 67
Jaubertia aucheri 75
Jean, Gerard 53, *54*
Jermy, Clive 104
Jodrell Laboratory (Kew) 155, 161
  collecting for 176–86
Johnson, Margaret 177, 178
  African expeditions 185
Jones, Professor Keith 177
Jordan River Gorge (Vanuatu) *124*
Juncus capitatus 82
Juniperus excelsa 84
Juniperus recurva *92*
Justicia hepperi 27

K

Kantega peak *92*
Kapachira *42*, 47
Kawozya peak 48
Keay, R W J 24
Kent, Sam 47
Kenton, Ann 177, 178, 183
Kenya expedition 184
Kerr, William 6–7, 15, 18
Kew Bulletin, 1914 19
  economic plants 147
Kew Guild 172
Kew Horticultural Diploma Course 170
Kew Magazine xiii
Kew students' travels and
  expeditions 170–5
Kharga Oasis (Egypt) *81*
Khumbu Glacier 94
Kielmeyera spp *137*
Kinchinjunga (Kangchenjunga) *12*
Kinabalu, Mt. 199, 201, *212*
Kingdon Ward, Frank 96
Kingdon Ward Prize 172
Kirkham, Anthony 167
Kniphofia 44
Kohautia 26
Kolombangara island (Solomon
  Islands) *120*
Korea, South *166*
Kotschy, Theodor 79
Kurawa (Kenya) *34*

L

Lactarius spp 206
lady's slipper orchid 126
lampshade poppy 102
Larix potaninii 101
Lasiurus scindus 76
legumes
  useful 147–61
Leguminosae 147
Lengwe Game Reserve 47
Leptadenia 30
Lepturus repens 71
Lewis, Gwilym 147–61
Lichiang (Lijiang) *97*, 99–101
Li-yun-chang, Professor 98
Lilies, ginger 95
Lilium henryi *17*, 19
Lilium lancifolium 7
Lilium souliei 102
Limosella 82
Linaria pelisseriana 82
Lippia rugosa *25*
Lithocarpus pachyphylla 101
Lithodora hispidula 82
Liverpool International Garden
  Festival 175
Livingstone Falls *see* Kapachira
Liwonde Game Reserve 48
Lloyd Binns, Professor Blodwen 42
Lobelia gibberoa 46
Lobelia mildbraedii 46
Lobelia stuhlmannii *45*
Lobelia trullifolia 46
Lobelia tupa 168, *169*
Lomatophyllum 66
Lomatophyllum aldabrense *67*, *70*
Lonchocarpus 148
Lonicera iaponica 6–7
Lonsdale, John 182
  Argentinian collections 166–7
Lord Howe Island *124*
Louvelia 53, 57
Lucas, Grenville 209–213
Ludlow, Frank 96
Lunt, William 19
Lupinus mutabilis 183
Lycium shawii 75

M

McBeath, Ron 98
Macrolobium acaciifolium 158
Madagascar 51–60
  expedition (1968) 52–60
  Masoala peninsula 53
  northeast, map **5**
Maerua crassifolia 75
Magnolia wilsonii 97
Malagasy bulbul 66
Malagasy Republic 51
Malagasy turtledove 66
Malawi 40–50
  map **41**
  University herbarium 48–9
Malaysia, Peninsular
  fern expeditions 199–201
  map **16**
Mali 28, 29
  seed collecting trip 192–5
Malus rockii 97
Mann, Gustav *14*
Manusela National Park (Indonesia) *201*
Maracá Rainforest Project 153–5
  map **13**
Marattia 196
Masson, Francis 2, 15, 16
Mayo, Simon J 128–43
Meconopsis delavayi 100
Meconopsis integrifolia 102
medicines 148
Medusagyne oppositifolia *69*
Meikle, R D 72, *82*
  Cyprus expedition 78–84
Mekong-Yangtse divide (China) 101–2
Melocactus smithii *153*, 154
Melville, Ronald *125*
Menzies, Archibald 4–5
Merton, L P H 79
mesembryanthemums 2
Mexico expeditions 176, 177–8
Milliken, William 155, 156
Milne-Redhead, Edgar 20
  correspondence to East Africa 31–7
Mimusops angel 190
minaret of Djam *87*
Mintirib 75
mistletoe 49
Mitchell, Robert 123, *173*
monkey-puzzle tree 5
Montagnea arenaria 77
Moore, Dr Hal 58, 177
Moringa 39
mouse-tail 83
Mujica, Professor 179
Mulanje cedar 44

Mulanje, Mount 43–6
Munthali, David 47
Musaceae 160
Museum of Economic Botany 145, 147
Myosurus minimus 83
Myricaria germanica 91

N

Napier Range (Western Australia) *125*
Natural History Museum (London) 104
Nandia domestica 7
Near East expeditions 72–84
Nelson, David 3
Neodypsis 53
Neodypsis lastelliana 57
Neophloga thiryana 53, 58
Nepal Himalaya 93–5, 213
Nepal map **8**
Nepenthes pervillei *68*
Nesogenes prostrata *67*
New Guinea 199–201
  see also Papua New Guinea
  fern expeditions
New Ireland (Papua New Guinea)
    106–8, 112–16
New Zealand
  Allan Cunningham's collections 9
  Ronald Melville's collections *125*
Niger 28, 29
  expedition (1841) 23
Niger Flora 23
Niger river 28, 29
Nigeria 28
  Charles Barter's collections 14
Nipa 109
nitrogen fixation 147
North Yemen *80*
Northumberland, Duke of 12
Nothofagus forest *109*, 110
nursery, temporary *114*
Nyika plateau *42, 44*, 46–8

O

oak, golden 79
oasis vegetation *76*
Ochna ciliata 67
Octomeles 113
Oldenlandia 26
Oldham, Richard 14–15, 18
Oleandra pistillaris *197*
Olluco *149*, 183
Oman
  Dhofar *81*
  expedition 72–7
  H M Sultan of 73
Oman Natural History Museum 74
Operation Raleigh 201, *201*
Ophioglossum 196
Ophioglossum nudicale *201*
Orania longisquama 53
orchids
  Andes collections 182–3
  of Papua New Guinea 110, 111, 115–
    16, 201
  Peru collections *171*
  of Solomon Islands 117–27, *173*
Orchids of the Solomon Islands 119, 127
Orchis italica *84*
Ormosia smithii 158
Ormosia 160
Ornithogalum chionophilum 83
Oryza longistaminata 195
Osmunda claytoniana 101–2
Owens, Simon 177
oxalis 2
Özhatay, Professor Neriman 178

P

Paeonia lutea *101*, 102
Paeonia mascula 83
Paepalanthus speciosus *129, 137*
palms 51
  carnaúba 139
  of Madagascar 57, 58
  of Sri Lanka 171
  of Thailand 172
Palomino, Guadelupe 178
Pandanus tectorius *67*
Panicum 70
Panicum laetum 195
Papaver postii 83
Paphiopedilium rothschildianum *212*
Paphiopedilum bougainvilleanum *121*, 122
Paphiopedilum wentworthianium 127
Papua New Guinea 103–16
  see also New Guinea
  expeditions 103–16
  map **10**
  rainforest *105*
Papyrus 30
Paraquilegia microphylla 100
Parkia pendula 160
Parris, Barbara, fern expeditions 199–
    202
Parrotia persica 163, *164*
Parthenocissus henryi 97
Passiflora 66
Patel, Hassam 43
Pattison, Graham 108
pau brasil 150, *151*
pelargoniums 2
Pemphis 66
Pennisetum spp 193, 194
Peru collections 183
Philippia 44
Philodendron insigne 139
Philodendron recurvifolium 136
Philodendron williamsii 136
Pholiota brittonae *205*
Photinia prionophylla 101
Phragmites 30
Picea farreri 97
Picea lichiangensis 100
Pico das Almas 139, 142–3
Pinguicula crystallina 83
Pinheiro, Raimundo 133
Pinus brutia *79, 82*
Pinus caribaea 180
Pinus patula 46
Piora ericoides 110
Piora, Mount *85, 109*, 110
pipewort *137*, 142
Piptanthus tomentosus 100
piranha 158
pitcher plant *141*
plant collecting and conservation 209–
    13
plant collectors
  dangers 18
  difficulties of work 16
  pay 16, 18
  qualities 15
Platycerium coronarium *197*
Plectranthus zebrarum 46
Plumbago aphylla 67
Poa langtangensis 95
Podocarpus 180
Podocarus milanjianus 28
Polhill, R M 49
  correspondence from East Africa 31–
    7
  journal of Fourth Colonial Office
      Expedition 38–40
Polunin, Oleg 93
Polyandrococos caudescens 134
Polycarpon 77
Polygala 27
Polypodium 196
Polystichum 196
Pometia 113
Pometia pinnata 104
Populus glauca 95
Populus szechuanica 100
Porter, Colin *175*
Prance, Ghillean T xi
Priest, Jim 163
Primula boothii var *alba* 95
Primula dryadifolia 100

Primula sherriffii 97
Primula sikkimensis 14
Primula verticillata *80*
Pringle, Sir John 2
Procryptocerus 159
Prosopis cineraria 73, 75
Protea angolensis var *trichanthera* 46
Protea spp 28
Providence, HMS 3, 4
Prunus himalaica 95
Psorospermum 28
Pteridium aquilinum 45
Pteridium aquilinum var wightianum *201*
Pteronia smutsii 19
Purdie, William 12–13
Puya *175*
Pycreus 70

Q

Quercus alnifolia 79
Quinoa *149*

R

Racemobambos nonohibernica 115
Radcliffe-Smith, Alan 95
rail *71*
rainforest *105*
  Brazilian *209*
  Guadalcanal *121*
  Maracá project 153, 154
  New Ireland *105*
Ranunculus 82
Ranunculus pseudolowii 110
Raphia palm 53
Ravenala madagascariensis *58*
Ravenea 57, 58
Rechinger, Karl 86
Remarkables (New Zealand) *125*
Renvoize S A *64*, 182
  Aldabra island expedition 61–71
restinga vegetation *132*, 134, 135
Réunion island *69*
Rhizobium 147
Rhododendron arboreum *92, 93*
Rhododendron arboreum subsp *cinnamomeum* var *cinnamomeum* *13*
Rhododendron decorum 98
Rhododendron ludlowii 97
Rhododendron microphyton 98
Rhododendron of Mekong-Yangtse divide 101
Rhododendron of Papua New Guinea 110
Rhododendron superbum, 116
Rhododendron wardii 97
Rhododendron yunnanense 98, *99*
rhododendrons, Joseph Hooker's collections 14
Rhododendrons of Sikkum-Himalaya *13*
riverine forest *56*
  Somalia *175*
Rock, Joseph 96
Rodgersia nepalensis 95
Roraima, Mount (Guyana) *140, 141*
Rosa banksiae 7
Rosa longicuspis 98
Rosa sericea *92*
Roscoea 100
Roscoea humeana *100*
rosewood 148
Rosularia sempervivum 166, *168*
rotenone 148
Rotman, Alicia 179
Royal Botanic Gardens, Kew
  responsibility for plant evaluation and distribution 162
  students 170
  travel scholarships 171–2
Royal Geographical Society
  Brazilian expedition 129
  expeditions 201
  Maracá Rainforest Project 154
  Wahiba Sands Project 72–7
Royal Horticultural Society Award of Merit 166
Royal Society 129
Rubiaceae 160
Rudall, Paula 178, 183
Russula 206
Rutherford, Sarah 171–2

S

Sabah map **16**
Sabaki river ferry (Kenya) *37*
St Mary's Hospital (London) 155
Salisbury, Richard 4
Salvia castanea 95
Sambucus megalophyllus 102
Sands, Martin J S, Papua New Guinea expeditions 103–16
Sandwith, Noel Y 19
Sarcopoterium spinosum *82*
Sarhad *90*
Saxifraga komorovii 91
Scaevola 66
Shebshi Mountains *24*, 26
Schilling, A D
  Daphne bholua collection 166
  Nepal expedition 93–5
  Sorbus microphylla collection 163
  Yunnan mountains expedition 96–102
Schönbrunn, Schloss *172*
Schweinfurthia papilionacea *76*
Scilla 86
Scilla morrisii 84
Sclerodactylon macrostachyum 71
screw-pine 67
Scrophularia chasmophila 100
Scutia 66
Seed Bank, Kew 177
  collecting 187–95
  containers *187*
  objectives 189–90
  wild collection 187–9
seed-collecting
  in China *100*
  expedition planning 190–2
Senegal 28
Seram *200*
Setaria 27
Seyani, Jameson 48, 50
Seychelles *68*
Sherriff, George 96
shipping plants 16, *18*, 19
Sibthorp, John 79
Sida 70
Sideroxylon 55
Silhouette Island (Seychelles) *68*
Simmons, John 163, 166
Sintenis, Paul 79
Sisyrinchium 178
Skimmia laureola var multinervia 95
slipper orchid *212*
Smith, Christopher 3
Smith, Gavin 172
Smuts, General Jan Christian 19
Sobralia *171*
Socotra island *68*
Solanum 66
Solmslaubachia pulcherrima 100
Solomon Islands
  map **11**
  orchid hunting expeditions 117–27, *173*
  tree seed collecting *191*
Somalia
  expedition 139, 190
  seed collection *189*
Sorbus microphylla *92*, 163
South Africa
  expedition 184–5
  Francis Masson's collections 2
  James Bowie's collections 8
  John Hutchinson's collections 19
Spain, collections from 19, 20
spore-prints 207
Sporobolus 71
Sporobolus virginicus 71
Sprekelia formosissima *180*

Spruce, Richard 128
Sri Lanka *see* palms
Stainton, Adam 93
Stapelia pulvinata *3*
Stathostelma pedunculatum 30
Stellera chamaejasme 100, 102
Stephanotis grandiflora 55
Stereosandra javanica 123
Stipagrostis plumosa 76
Stipagrostis sokotrana 76
strawberry tree 82
Streptocarpus 44
Striga 26
Studley College Trust 174
Stylosanthes 190
Styphelia suaveolens 110
Sulphur, HMS 11
Sultan Qaboos University (Muscat) 76
sunbird 66
Survey of Economic Plants for Arid and Semi-Arid Lands 190
Swallow, HMS 14
Swartzia grandiflora 158
Syagarus harleyi 138
Symbegonia *109*
Synge, Hugh 47

T

Tanzania Central plateau *40*
Taraxacum aphrogenes 84
tarwi 183
Taylor, Sir George 28–31, 94
  Kew Horticultural Diploma Course 170
Taylor, P 33
Tephrosia 26, 70, 148
Terminalia 27, 55
Terminalia brassii 104
Termitomyces 206
Termitomyces eurrhizus *207*
Thailand *see* palms
Thelypteris 196
Thesiger, Wilfred 75
Thesium 28
Thespesia populnea 67
Thompson, Dr Peter 187–8
threatened species 210
Tiger Lily 7
Tigridia ehrenbergii *176*
Tigridieae 178
Tilia likiangensis 100
toadflax 82
tortoise, giant *61*, 67, 70–1
Touré, Aboubacar 193
Townsend, C C 72, *80*
Tradescantia laxiflora 177
Trans-African Hovercraft Expedition 28–31
Travel scholarships from Kew 171–2
Trichocereus pasacana *181*
Trichomanes 196
Trichomanes pallidum 200
Trimezia martii 180
Trinidad *205*
Trochocarpa 110
Troödos mountains *82*
Tropidia disticha 123
Tsuga yunnanensis 101
Tulipa 86
Tulipa banuensis 89
Tulipa cypria 82
Tulipa kolpakowskiana 88
tunadi *159*
Turkey
  bulbous plants *81*
  expeditions 178

U

Uapaca 27
Ullucus tuberosus *149*, 183
Urginea 28
Utricularia 26

V

Vaccinium fragile 98
Vancouver, Captain George 4–5
Vanuatu people *124*
Vernonia kawozyensis 48
Vernonia nigritiana *25*
Vigna 195
Viola maymanica 90
Voanioala 51
Voanioala gerardii *57*, 60
Vogel Peak *24*, 26, 27–8
Vogel, Theodor 23
Vonitra 53
Vonitra fibrosa 57

W

Wadi al Batha 75
Wadi Andam 76
Wadi Matam 76
Wahiba Sands of Oman 72–7
  high dunes *73*
  low dunes *74*
Wakehurst Place 166
  Seed Bank 187
Wakhan Corridor (Afghanistan) 85–91, *90–1*
Waldheimia glabra 91
Waldheimia nivea 91
Wallich, N 93
Wardian cases 16, *18*, 19
water crowfoot 82
Wendelbo, Per 86
West Africa 23–31
  map **15**
West Indies, Francis Masson's collections 2
Wickam, Henry 128
Widdringtonia nodiflora 44
Wiles, James 3
Wilford, Charles 14
Wilkinson, Sue 19
Wilmot-Dear, Melanie 182
Wilson, Ernest Henry 96
Wood, Jeffery 108
World Wide Fund for Nature 137
Wye College (Kent) Malawi expedition 47–8

X

Xerophyta 44
Xyridaceae 142
Xyris 26

Y

yak *91*
Yemen *80*
Yuan-lui-kun 98
Yulong Shan (China) *97*
Yunnan mountains (China) 96–102

Z

Zaire, seed collecting *188*
Ziziphus mauritania 195

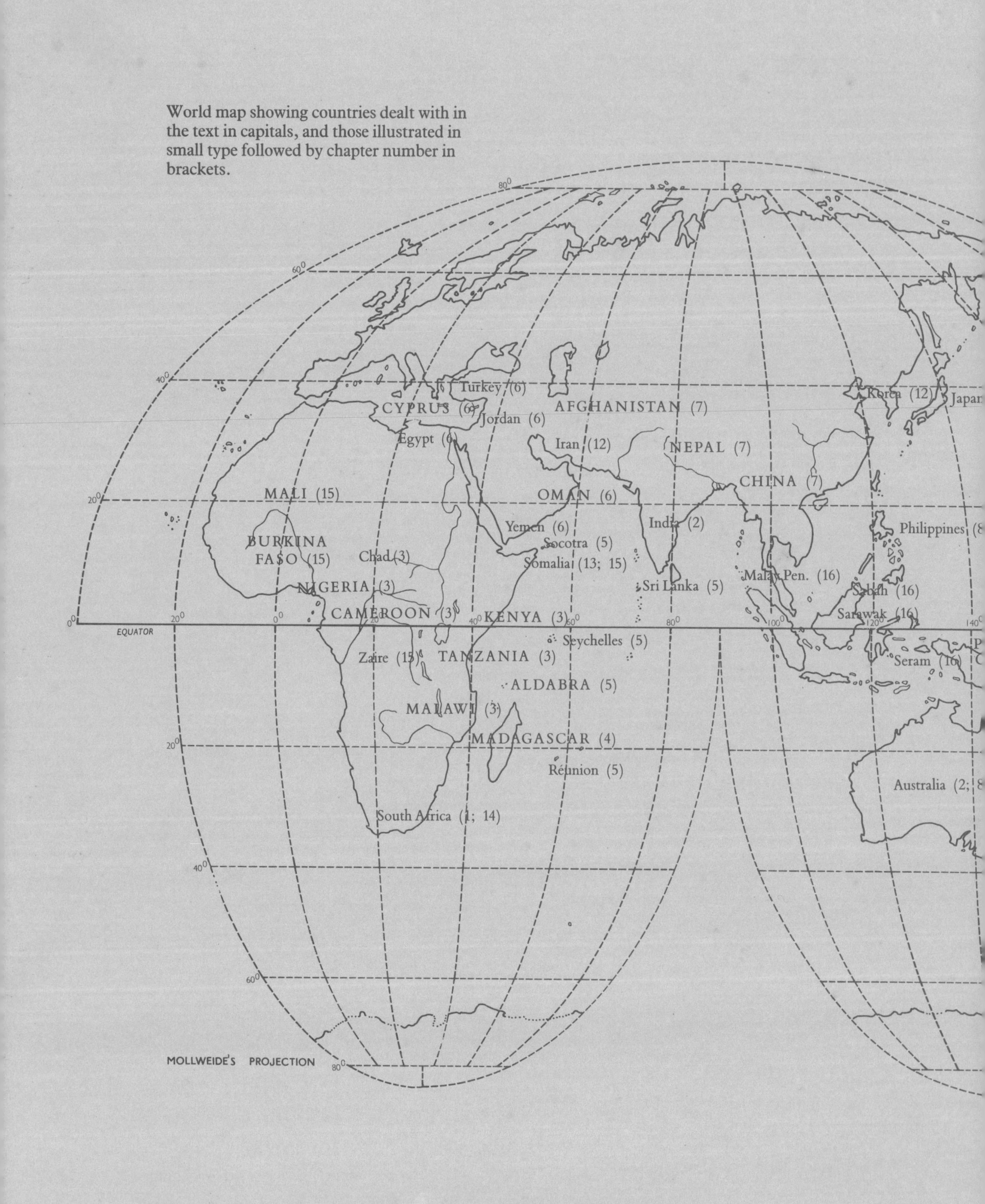

World map showing countries dealt with in the text in capitals, and those illustrated in small type followed by chapter number in brackets.